WATER QUALITY MODELLING

Water Quality Modelling

Edited by

Roger A. Falconer
Department of Civil Engineering
University of Bradford

Published in Association with
The Institution of Water and Environmental Management

Routledge
Taylor & Francis Group

LONDON AND NEW YORK

First published 1992 by Ashgate Publishing

Reissued 2018 by Routledge
2 Park Square, Milton Park, Abingdon, Oxon, OX14 4RN
52 Vanderbilt Avenue, New York, NY 10017

Routledge is an imprint of the Taylor & Francis Group, an informa business

Publisher's Note
The publisher has gone to great lengths to ensure the quality of this reprint but points out that some imperfections in the original copies may be apparent.

Disclaimer
The publisher has made every effort to trace copyright holders and welcomes correspondence from those they have been unable to contact.

A Library of Congress record exists under LC control number:

Typeset in 11 point Baskerville by Photoprint, Torquay, Devon

ISBN 13: 978-1-138-35220-9 (hbk)
ISBN 13: 978-1-138-35223-0 (pbk)
ISBN 13: 978-0-429-43248-4 (ebk)

Contents

List of contributors   vii

Editor's preface   ix
Roger A. Falconer

Foreword   xi
K. Guiver

1   Water quality modelling: An overview of requirements   1
    R.J. Pentreath

2   Parallel processes in hydrology and water quality:
    Objective inference from hydrological data   10
    Peter C. Young

3   The role of models in environmental impact assessment   53
    Paul Whitehead

4   Modelling river water quality and impact from sewers
    and storm sewer overflows   69
    Anders Malmgren-Hansen and Hanne K. Bach

5   Research developments of flow and water quality
    modelling in coastal and estuarine waters   81
    Roger A. Falconer

6 Hydrodynamic and physical considerations
   for water quality modelling   110
   Mrs J.M. Maskell

7 Water quality aspects of estuary modelling   119
   J.I. Baird and K. Whitelaw

8 Mathematical models and engineering design   127
   Graham Thompson

List of contributors

Hanne K. Bach Ecological Modelling Centre, Danish Hydraulic Institute and Water Quality Institute.

J.I. Baird WRc Environment.

Roger A. Falconer Professor of Water Engineering, Department of Civil Engineering, University of Bradford.

Anders Malmgren-Hansen Ecological Modelling Centre, Danish Hydraulic Institute and Water Quality Institute.

Mrs J.M. Maskell Manager of Environmental Studies, Hydraulics Research Ltd.

R.J. Pentreath Chief Scientist, National Rivers Authority.

Graham Thompson Divisional Director, Binnie and Partners.

Paul Whitehead Head of Environmental Hydrology Division, Institute of Hydrology.

K. Whitelaw WRc Environment.

Peter C. Young Centre for Research on Environmental Systems, Institute of Environmental and Biological Sciences, University of Lancaster.

Editor's preface

Roger A. Falconer

In recent years increasing emphasis has been focused on the use of numerical or computer models for predicting flow fields and water quality characteristics in rivers and sewers, and coastal and estuarine waters. Such models are now being used widely by consulting engineers and water companies for environmental impact assessment studies and design considerations, both for UK and overseas projects. With the added benefit of significant developments in high quality colour graphics, considerable advances have recently been made in water quality modelling and it was therefore opportune for the IWEM Tyne and Humber Branch to organize and host a one day symposium on this subject. The meeting was held in Harrogate on 13 November 1991 and was attended by over 120 delegates.

The symposium was opened by the President of the Institution, namely Mr Ken Guiver, MBE, whose opening remarks set the scene for the meeting and are included in this book. Dr Jan Pentreath, Chief Scientist of the National Rivers Authority, gave the opening keynote paper on an 'Overview of Modelling Requirements'. Dr Pentreath was unable to attend the symposium on the day since, at short notice, he was asked to join HRH The Prince of Wales at a meeting elsewhere in Europe and Mr Mervyn Bramley gave an excellent presentation of Dr Pentreath's paper in his absence. The remainder of the symposium was split into two sessions, namely on River and Sewer Modelling and Coastal and Estuarine Modelling respectively.

The first session started off with a presentation of the Aggregated Dead Zone Model by Professor Peter Young, of Lancaster University, and was followed by presentations on the MAGIC and QUASAR models for EIA river studies by Dr Paul Whitehead of the Institute of Hydrology, and an overview of river and sewer and storm sewer water quality modelling by Dr Anders Malmgren-Hansen of the Danish Hydraulic Institute.

The session on Coastal and Estuarine Modelling opened with a presentation on some current research developments in the field relating to turbulence modelling, nested and patched modelling and refined and higher order schemes; this was followed by presentations by Mrs Jackie Maskell, of HR Wallingford Ltd, on the ability to simulate all the relevant hydrodynamic processes and Dr Jim Baird of WRc, on general aspects of estuarine water quality modelling. Mr Graham Thompson of Binnie and Partners, explained how models were used in engineering design for managing the environment and Mr Philip Nicholas of Wallace Evans highlighted the requirements of water companies for model use. The meeting was then summed up by Dr Tony Edwards, Regional Manager (Environment) of the National Rivers Authority, Yorkshire Region.

In the organization and planning of the symposium leading to this book I am grateful to all colleagues of the IWEM Tyne and Humber Branch Committee for their support and encouragement and, in particular, to my colleagues on the organizing committee, including:

Mr Hugh Brooksbank, Consultant, Bullen and Partners
Dr Tony Edwards, Regional Manager (Environment), NRA Yorkshire Region
Dr John Hudson, Area Manager (Calder), Yorkshire Water Services Ltd

In the preparation of this volume I am also grateful to the authors, my secretary, Christine Dove, IWEM headquarters staff, including Patricia Walters and Lavinia Gittins, and Ashgate's Consultant Publisher, John Hindley.

Roger Falconer
Bradford 1992

Foreword

K. Guiver

The subject of water quality modelling is important at the present time in view of the emphasis on improving environmental standards of rivers and coastal waters, whilst at the same time ensuring that the expenditure on improvements gives value for money. This book is based on contributions to an IWEM symposium held on 13 November 1991 at Harrogate, organized by the Tyne and Humber branch of the institution under its Chairman Professor Roger Falconer.

It might be said that the first water quality model ever used in the UK was that devised by the Royal Commission on Environmental Pollution in 1912, with its recommendation that biological purification of sewage can give rise to satisfactory river water quality provided there is always at least an eight to one dilution for any such discharge. This guidance was based on the fact that biological purification could produce an effluent of 20 mg per litre biochemical oxygen demand (BOD), and that dilution of eight to one with clean riverwater of a BOD no greater than 2 mg per litre meant that the resulting mixture had a BOD no greater than 4 mg per litre which was still quite satisfactory.

Tremendous developments have taken place since that time, both in purification practices and in prediction of changes that can be achieved, using mathematical models. In particular, with the increased power of computers in recent years, there has been a major step forward. New models have been devised to deal with the chemistry and biology of rivers and estuaries taking into account flows, sediment loads etc., under the various titles such as TOMCAT, MAGIC, RIVPACS and MOSQUITO etc. With the wide use of these abbreviations, the subject of modelling has developed a language of its own. That is not to say however, that the subject has become bedevilled by jargon. Computers using the models are user friendly, as they must be if they are going to be widely understood and used not only by computer experts, but engineers and scientists, geographers and mathematicians in the industry.

Part of the pressure for the development of modelling in the UK has come from the privatization of the water industry and the formation of the new regulator, the National Rivers Authority. This has given rise to improvement in the knowledge of processes and the ability to predict what might happen to the water environment under changing circumstances. River and estuarine flow patterns can be assessed, together with their influence on water quality. Weather forecasting and rainfall itself which is a fundamental of the hydrological cycle can be more accurate with the use of models and the power of computers.

It is not only in the UK that the research work has been carried out for the surge forward in predictive capacity. The chapters presented here include an important contribution from Dr Malmgren-Hansen from Denmark.

There is a wide interest in quality modelling throughout Europe, particularly in those countries bordering the North Sea and Baltic Sea, both of which suffer severely from pollution, and where major improvements are being implemented for many discharges.

It is not only in helping to obtain the best answers to environmental pollution problems that water quality models are of great value. They can be also helpful in demonstrating the solutions at meetings and public enquiries. Use of colour graphics can demonstrate successive patterns of tidal ranges and the changes that occur in tidal waters from coastal discharges. The models used can take into account wind direction and speed and the likely death rate of organisms present in the discharges, dependent on factors such as light intensity and sunlight. In a day and age when there is criticism of some anti-pollution measures, and some cynicism of expert advice, the use of water quality models must be seen as a part of the professional approach to finding the best answers to environmental pollution problems.

K. Guiver, President
The Institution of Water and
Environmental Management

1 Water quality modelling: An overview of requirements

R.J. Pentreath

Introduction

The NRA uses models extensively in support of all of its principal areas of responsibility, and none more so than in the field of water quality. Virtually all of these models had been developed prior to the NRA's formation, and thus prior to the 1989 Water Act and the later Environmental Protection Act of 1990. Such models had served the previous water authorities well, and many will continue to serve well the needs of the NRA. But needs must change and the NRA has very specific requirements arising from the legislation which set it up. It also has to consider the implications of being a national body, and the pressures to be demonstrably cost-effective in its operations. Thus before discussing any specific modelling requirements, it is worth reviewing what exactly the NRA has to do within the field of water quality.

Responsibilities with regard to water quality

Both the Water Act 1989 and the Environmental Protection Act 1990 place a number of duties on the NRA, and give it various powers. Collectively these require and enable the NRA to maintain and improve where necessary the quality of inland, underground, estuarine and coastal waters out to a distance of 3 nautical miles from the shore. The NRA has also to carry out the work required to comply with a number of

EC Directives and other international obligations which, together with a system of water quality objectives, form the framework within which the NRA's water quality functions operate. In essence, therefore these require the NRA to:

- advise the Secretary of State with regard to the setting of statutory water quality objectives;
- review, set and, where necessary, revise consents for discharge in order to meet such objectives;
- demonstrate compliance or otherwise with discharge consents;
- take enforcement action with regard to breaches of consent where appropriate, and for other acts of pollution;
- remedy or mitigate the effects of pollution where it does occur, and of preventing it at source;
- monitor the extent of pollution in the environment and, through such programmes, demonstrate compliance with statutory water quality objectives;
- and all within a general framework of conserving and enhancing the natural beauty, and amenity, of inland and coastal waters.

Setting water quality objectives

The 1989 Water Act allows for the setting of water quality objectives by the Secretary of State: once set, it is a duty upon both the Secretary of State and the NRA to ensure that such objectives are met. The precise form of such setting has yet to be fully worked out, although certain existing EC Directives are being introduced via the relevant water quality objective sections of the Water Act. There are several modelling aspects here: determining how use-related objectives can be met within catchments, tuning consent setting as required, and demonstrating the extent of compliance.

At present NRA regions use river models such as QUASAR, plus SIMCAT and TOMCAT, for such purposes. These models provide a practical means of setting consents, and of assessing the achievement and compliance of water quality objectives. They will therefore continue to be used for some time; but the future should demand much more. It is self-evident that there is a limit to the amount of potentially polluting material that a catchment area can accept without causing a deterioration in the quality of the water or having longer-term adverse effects. It is therefore no longer acceptable simply to consent discharges on the basis of an assessment of the impact of the cumulative upstream discharge at a particular point, by reference to the concentration of certain substances. The fate of persistent chemicals which may be present will also have to be taken into account; these chemicals are often associated with particulate materials and may accumulate elsewhere.

Because the NRA is responsible for water quality in fresh waters, estuaries, and the first three miles of the sea, it needs to be accountable for the fate of those substances the discharge of which it has consented. The problem here is in striking the right balance. Aspects which are clearly difficult are the modelling of:

- adsorption/desorption of chemicals at the freshwater/marine interface;
- particle settlement and re-mobilization within estuaries; and
- the interface of such processes with those occurring in the river and the offshore zones.

All of this builds an image of immensely complicated site specific models, which may gladden the contract modeller but is likely to remain no more than an ideal objective. On the other hand, it will become increasingly difficult to pursue policies in the future which may result in the expenditure of large sums of money to reduce the input of specific chemicals, without putting a considerable effort into determining the likely effectiveness of such measures in advance, particularly in the context of the undoubted arguments which will arise over point source and diffuse source pollution. Such arguments are bound to develop over the discharge of nutrients, some metals, and pesticides. A first step might therefore be to develop some fairly simple generalized models of a complete system – river, estuary, plus coastal water 'box' – with the basic parameterization of the hydraulic features, particulate material transfer, variable sorption kinetics and so on, simply to obtain a better perspective of such issues for our major river systems. It might also be useful to build around existing models, with a statistical presentation for the analysis of time series data. In the longer term, however, the exploitation of GIS technology plus modelling may provide the most powerful tool for large scale catchment management.

The setting of water quality objectives will of course be preceded by considerable discussion and consultation, the driving force behind them being an assessment of the uses to which the water is, or could be, put. Where an improvement is necessary, discussion will centre around the benefits to be achieved and the effort required to achieve them. In the first instance this will be a subject not so much of modelling – which ultimately it should be – but of less sophisticated means of handling the data. But once set, models will be used to adjust consented discharges in order to obtain the quality objectives: the aim here should be to obtain the most cost-effective solution.

For urban waste water discharges an entire menagerie of models exists, such as MOSQITO, QUALSOC, CARP and the recently enhanced MIKE 11: such models and their use – particularly in relation to storm sewer overflows – are being progressed by the Urban Pollution Management Programme, which the NRA supports via the Foundation for Water Research. But the extent to which the cost-effectiveness of any solution is optimized is less clear: in other words, whilst it can be shown that water quality objectives could be met in a specific area by several combinations of discharge control, it is not always clear which would be the most cost-effective when different dischargers are involved. Where an objective has to be met primarily by the interplay of several consented discharges from the same discharger, such as a water services company, then no doubt the discharger will give serious consideration to obtaining the optimum solution. But the NRA will also have to consider the relative effects of different dischargers, for which the data may well be more difficult to obtain and evaluate. Early attempts to do so will inevitably be unsophisticated.

The cost implications of achieving improved water quality by the setting of water quality objectives will become increasingly important. The 1989 Water Act allows for the NRA to request that the objectives, once set and attained, be revised. If one views these objectives as a set of stepping stones towards achieving improved water quality, each step being of some 3 to 5 years duration, then each subsequent step will be harder to take. And increasingly the onus will be on the NRA to argue the case of the benefits to be gained from such improvements, or count the cost of a failure to do so, and balance them against the estimated cost of improvements. This will involve some degree of economic modelling.

The greatest scope for economic modelling, however, is to some extent denied to the NRA because of the arrangements in the Water Act. Under the Act, the NRA is able to recover its costs for the work it needs to do to consent discharges and carry out the accompanying compliance monitoring programme. The Act does not allow for a system of *incentive* charging with regard to discharge consents although, no doubt, such an allowance will eventually materialize. This would give considerable scope for the economic modelling of the short- and long-term consequences of different ways in which a combination of statutory limits and incentive charging could be applied – to effluent dischargers, to abstractors, to the producers, retailers and users of chemicals, to land developers and, not least, to those who would stand to gain by, or be prepared to pay for, the benefits. Indeed, it is essential to begin such modelling now.

Away from the rivers

By their very nature, estuarine and coastal water models are difficult to construct. There are several 1-D models in use within the NRA which are specific for certain estuaries – the Medway, Stour, Thames Tideway, Mersey, Humber, Severn. A number of 2-D models are also available for selected estuaries, or parts of an estuary, and for some coastal waters. The estuarine models have usually been developed in connection with a specific form of discharge, such as that from the titanium dioxide industries in the Humber for which the need essentially arose from an EC Directive. The depth-averaged 2-D coastal models have been developed to evaluate the design and location of sea outfalls for sewage.

At this stage it is difficult to argue for the short-term requirement of better coastal models for such planning purposes; other requirements are likely to arise from the setting of water quality objectives, however, and from the needs to account further for particle reactive chemicals entering coastal waters. Unfortunately there has been little attempt to interface estuarine and coastal water models, or near-shore models with those used offshore – in the North Sea context, for example. Such models are also expensive and require substantial data bases.

Groundwater models, as one might expect, are primarily driven by water resource requirements. Some – such as FEPOLL, AQUA and MOC – address groundwater flow and contaminant transport in order to define the measures needed to restore or protect groundwater quality. The future needs here, however, are considerable. As water resources become increasingly under pressure, the need to protect underground supplies, and to clean them up where necessary, will increase. Two pressures in particular need addressing. Whatever the arguments over the sensibility of standards for nitrate in drinking water, the recent EC Directive on nitrates demands that greater attention be paid to such contamination. The potential costs and effects are substantial, such that some investment in modelling is essential. 'Mixing cell' modelling has been used to calculate nitrate concentrations in response to changes in land use, but this needs to be expanded into full 3-D for selected areas, and a representation of nitrate leaching incorporated.

The second pressure is that arising from the proposed EC landfill waste directive and the need to advise and react to proposed, existing, and developed sites. The scale of the problem is quite daunting and a fundamental rethink of both the modelling and data base requirements is likely to emerge over the next year or so.

Operational modelling

Simulating the effects of pollution incidents in rivers is a necessary input to catchment planning; indeed 'planning' models such as QUASAR and MIKE 11 are used for such purposes. For major river systems it is highly desirable to develop such models such that they have 'real time' capability, linked to telemetered data derived from stations situated throughout the catchment. The Thames is already 'wired up' in such a way and this trend will continue as money and experience permit. But pollution incidents are not the only concern. The NRA also has an interest in modelling the effects of pollutants; for example, what happens to pesticides after their application to land? Such models need to take account of the many parameters which affect the leaching of pesticides into ground waters or surface waters.

Process modelling

Without a reasonable understanding of environmental processes it is very difficult to make sensible decisions on pollution control. The need and utility of process modelling in relation to water quality has been amply demonstrated with respect to the effects of acid rain. Abatement of such pollution involves a very large expenditure, and it is essential to have a good understanding of the consequences of the measures taken. The potential to waste large sums of money in relation to the belief that riverine inputs of nitrates can effect eutrophication in estuarine and coastal waters is considerable, unless a better understanding is obtained soon of the balance of nitrate input and denitrification in such areas. Such modelling is being developed, supported by a co-ordinated research programme.

A similar problem exists with respect to the presence of blue-green algae in inland waters. Expensive attempts to remove phosphate levels in certain areas in order to reduce the occurrence of such algae may not be appropriate in the majority of cases. The NRA's review of blue-green algae recognized that a number of management options should be available in order to reduce or prevent the algae blooming, depending on the characteristics of the water. The key processes have been modelled and a management model is being developed which can be applied to any body of water. Its effectiveness will depend not only on the data bases available, but on the accuracy and understanding of the processes being modelled.

Model size and scale

Getting a model's size and scale right for the job is one of the most difficult modelling skills. There is often a tendency to be over-ambitious, such that a modeller's skill and enthusiasm soon outgrows the data bases and processor time available. Models often never get completed or, if they are, can only be operated by a few. Models without comprehensive manuals are of no value to organizations such as the NRA; and writing models today which are not 'user friendly' is inexcusable.

Although the number of models generally available is always expanding, the number used by the NRA is relatively small – for the reasons already mentioned. With a greater emphasis on catchment accountability, and involvement in estuaries and the coastal zone, demands for 3-D modelling will grow. There is, of course, considerable scope for 'plus a half' models because all too often it is the case that most of the behaviour of a system can be described by one or two dimensional modelling as appropriate, such that the second or third dimension respectively can be approximated by a cruder representation. Such 1½-D and 2½-D models have been extensively used in the marine field in relation to, for example, examining the effects of particle scavenging on the distribution of non-conservative substances.

The size and scale of models is also linked to the development of processing power and the organization of data bases. The incorporation of parallel processors into less expensive micro- and mini-computers should enable more comprehensive models, and real time simulation, to be achieved. The changing nature of large data sets, such as soil survey classifications and digital terrain mapping, provide spatial information patterns as opposed to 'point' data sets; current models cannot make direct use of such information. This is, however, a rapidly expanding field and it appears that models are being developed in the risk assessment field – risks of pollution, and of eutrophication – which make use of such data bases.

Data sets are also needed to assist in the interpretation of model outputs, such as the potential for a comprehensive toxicological data base. Thus model outputs currently in the form of concentrations of substances in water would, instead, use 'ecotox' data bases to provide direct information on the potential aquatic 'health' of the system.

Calibration and validation of models is essential: it is therefore surprising that such little thought is given to the production of useful adjuncts to the modelling effort, such as procedural handbooks and data

7

sets of ranges of determinand values. Standard rate constants for nitrification and denitrification processes, K_d values, biological concentration factors, biological half-times, T_{90} values for bacteria, critical wind speeds for plume dispersion models and so on, all need to be standardized in an organization such as the NRA; variances in such information used – usually at the whim of the operator – is often the cause of much misunderstanding. In fact there is a growing need for what might best be described as bench testing, auditing, and the harmonization of existing models in order to achieve national standards of their performance. If models and their data bases are not adequately quality controlled, nationally, then what degree of reliance can be placed on the decisions that are made across the country which are dependant upon their use?

Monitoring and modelling

The development of modelling techniques, like all of the NRA's R&D activities, needs to be aimed at making the NRA more effective and efficient in the tasks which it has to carry out. The effectiveness of the NRA in turn will largely be dependent on the extent to which good environmental conditions, and improvements where necessary, can be clearly demonstrated. This requires monitoring and surveillance programmes.

First of all, the NRA needs to ensure that its monitoring programmes are efficient; research has been commissioned to model the optimum amount of sampling required to demonstrate compliance with given standards and criteria, and to detect significant trends. Secondly, extensive model testing has been carried out since the NRA was formed of the role of biological data in classifying river systems and how they can be used to reduce the risk of wrongly representing the state of a river on the basis of physical and chemical data alone. Depending on the success of such testing, further model evaluation will be carried out to see to what extent the approach could be applied to other types of surface water. Such models as RIVPACS are, however, very demanding of large data sets in order to set them up.

There is also a larger role, however, for the use of models in developing strategies, because monitoring is essentially a game. Models based on game theory, therefore, are appropriate. The nature of the games may differ. The objective of a monitoring strategy may be to demonstrate that a certain environmental condition complies with a desired target condition – for example, that a given stretch of water complies with a required water quality objective. Alternatively, the objective of such a strategy may be to demonstrate that a certain discharger is not attempting to comply with a discharge consent over a given period of time. In each case resources to monitor are limited, and the game is played based on certain set rules, experience and strategy. As a result of the Water Act, for example, evidence of breaches in discharge consents has to be provided by taking 'tripartite' samples, which are much more expensive than normal monitoring samples. The frequency of tripartite sampling to which a discharger may be subjected might, therefore, depend on his record of compliance in an attempt to ensure that the discharger will choose to act differently. Such a monitoring strategy would aim to minimize the expected number of breaches of consent to a certain number within a given population of dischargers. Other strategies need to be optimized such that expected environmental standards are met.

There is clearly much scope for a combination of game theory and economic modelling in this field which does not appear to have been significantly developed in the UK. Of course there are other ways of playing the game: the NRA's R&D work is also directed towards the production of better instrumentation for monitoring compliance or otherwise of discharges and receiving waters. Such instruments should make tripartite sampling more efficient, and ensure that vigilance is maintained at all times where necessary.

Discussion

It is easy to construct shopping lists of models, as it is for other R&D requirements. It is less easy to derive common priorities and time scales. For the NRA, however, it is clear that emphasis must be given to the modelling support required to ensure that water quality objectives are properly set and that, once set, they are complied with; models to ensure that such objectives are met, and compliance demonstrated, in the most cost effective manner are also essential over the next two or three years.

The more complex catchment models, and economic and game-theory models, each have their place. It is, however, likely that questions to which such models could provide answers will increase considerably over the next two or three years; although it would be a mistake to expect too much too soon. Nevertheless, the more adventurous should give serious thought to such priorities now.

2 Parallel processes in hydrology and water quality: Objective inference from hydrological data

Peter C. Young

Introduction

Over the last few decades a controversy has arisen between some environmental scientists about the way in which models of environmental systems should be formulated. On the one hand, there are those who feel that mathematical and computer-based models should reflect the perceived complexity of environmental systems and should have a structure which resembles closely the physical, biological and chemical structure of the real world *as it is understood by them*, given the contemporary state of scientific knowledge about the system. In such *simulation models*, the equations, usually in the form of ordinary or partial differential equations, are obtained in various ways, e.g. from dynamic conservation equations (mass, energy, momentum, etc.), which have a direct physical interpretation. On the other hand, scientists with a statistical turn of mind tend to warn of the dangers inherent in such a 'mechanistic' or 'simulation modelling' approach and favour 'data-based' procedures, where the model structure is inferred, and the model parameters are estimated, by reference to experimental data using more objective, statistically based methods. Following from such considerations, the physically-based (or mechanistic) models are normally fairly complex in structure, deterministic in form, and are characterized by many parameters. In contrast, the data-based models are often simple in structure, inherently stochastic in form, and are characterized by only as many parameters as can be justified by the information content of the available data (i.e. they are 'parametrically efficient' or 'parsimonious'; see Box and Jenkins, 1970).

As in most controversies, however, the lines of argument are drawn too rigidly and the way in which environmental models are constructed in practice is rarely as clearly defined as the debate would suggest. For example, if the available data are severely limited, then the more speculative mechanistic models become an almost essential extension of the 'mental models' of the scientist. Moreover, when accompanied by exercises in sensitivity analysis, they can prove extremely useful in examining ideas about alternative mechanisms and their relative importance in defining the behaviour of the model system. At the other extreme, where plentiful supplies of data are available, even the truly 'black box' time-series models can prove very useful in understanding more about the nature of the data and allowing for exercises such as seasonal adjustment, interpolation over gaps in the data, and forecasting.

But there is an intermediate type of model which we discuss in this chapter; one which exploits the availability of time-series data in statistical terms but which overtly attempts to produce models which, although they are much more parametrically efficient than the normal mechanistic model, have a sensible physical interpretation. In these *Data-Based, Mechanistic Models*, the structure is obtained by some form of objective statistical inference applied to a given, general class of time-series model, but the resulting model is only considered fully acceptable if, in addition to explaining the data well, it also provides a description which has relevance to the physical reality of the system under study.

The Transfer Function (*TF*) model, in discrete or continuous-time form, provides a particularly useful, general tool for such data-based mechanistic modelling and it has been exploited in various areas of environmental systems analysis, such as hydrology (e.g. Young and Wallis, 1985; Young, 1986; Young and Beven, 1992), soil science (e.g. Beven and Young, 1988) and biology (e.g. Gould et al, 1988; Young and Minchin, 1991), during the last few years. In this chapter, we present a very general approach to *TF* modelling and show how it can provide a powerful tool for data-based mechanistic modelling applied to the investigation of parallel processes in hydrology and water quality systems.

The importance of parallel processes in hydrological systems is obvious: in almost all natural hydrological systems, water can travel through a variety of pathways between its origin as rainfall and its eventual arrival at the sea. Some parallel pathways are obvious because we observe them directly; as when rainfall passes through small streams in a catchment before it reaches the main river channel; or when a river

11

passes either side of an island or through a cave system containing multiple channel and tunnel systems. Other parallel processes, such as those which occur in the inaccessible subsurface and ground water are, however, much more obscure since we cannot observe them directly and need to rely on inferring their presence from rainfall and streamflow measurements made on the surface.

In the case of a river catchment, the most popular approach to solving this inferential problem is based on the construction of simulation models. As pointed out above, these usually reflect the model builder's perception of the physical system, in this case the parallel structure, based on reference to the available data and any *a priori* information which may provide insight into the physical nature of the catchment. In this chapter, we do not question this pragmatic approach to modelling, which often yields very useful results. But we do suggest that, if reasonable quantities of time-series data are available on the catchment, then these should be exploited fully *prior* to simulation modelling, using the powerful procedures for *TF* modelling discussed in subsequent sections of the chapter. We also show how the same *TF* modelling procedures can be useful in a laboratory situation, by considering its application to the analysis of breakthrough curve data obtained from tracer experiments on soil columns.

Transfer function models

It is possible to unify *data-based* model analysis in terms of a general operator ρ and represent the relationship between an input time-series $u(j)$ and an output series $x(j)$ by a *TF* model of the following general form,

$$x(j) = \frac{B(\rho)}{A(\rho)} \, u(j-d) \tag{1}$$

where $A(\rho)$ and $B(\rho)$ are polynomials in ρ. In practical applied systems analysis, the most popular special cases of this model are:

(1) **The continuous-time (time-derivative or Laplace operator) *TF*.** This is probably still the most popular *TF* form for conventional control systems analysis, where:

j is replaced by time t

d is replaced by τ, the pure time delay in time units

ρ is replaced by time derivative operators $s = \dfrac{d}{dt}$

(2) **The special discrete-time (delta operator) *TF*.**
This is a discrete-time approximation to the continuous time *TF* where:
j is replaced by *k*, the sampling integer
d is replaced by Δ, the pure time delay in sampling intervals
ρ is replaced by the delta operator δ

(3) **The standard discrete-time (backward shift operator) *TF*.**
This is the most popular *TF* form for identification and estimation studies, where:
j is replaced by *k*, the sampling integer
d is replaced by Δ, the pure time delay in sampling intervals
ρ is replaced by the backward shift operator z^{-1}

Most experimental and monitored time-series obtained from hydrological systems are in discrete-time, sampled data form and are analysed in the digital computer. For this reason, we will concentrate here on the two, major discrete-time *TF* model forms (2) and (3). Note, however, that the δ operator *TF* model (3) is the discrete-time equivalent of the Laplace operator *TF* model and is a logical replacement for that model in a digital context. In addition, continuous-time *TF* models can be estimated (see Young and Jakeman, 1980; Young, 1981) using similar procedures to those used for δ operator estimation.

Backward shift (z^{-1}) TF model

The general, discrete-time, z^{-1} operator *TF* representation of an *nth* order single input, single output (*SISO*), discrete-time system, with a sampling interval of Δt time units, is normally written in the following form,

$$x(k) = \frac{B(z^{-1})}{A(z^{-1})} u(k-\Delta) \qquad (2)$$

where z^{-i} is the backward shift operator, i.e.,

$$z^{-i}x(k) = x(k-i)$$

and $A(z^{-1})$ and $B(z^{-1})$ are the following polynomials in z^{-1},

$$A(z^{-1}) = 1 + a_1 z^{-1} + \ldots + a_n z^{-n}$$

$$B(z^{-1}) = b_0 + b_1 z^{-1} + b_2 z^{-2} + \ldots + b_m z^{-m}$$

In general, no prior assumptions are made about the nature of the transfer function $B(z^{-1})/A(z^{-1})$, which may be marginally stable, unstable, or possess non-minimum phase characteristics.

The discrete differential (δ) operator TF model

An interesting alternative to the z^{-1} operator *TF* model is the following 'discrete differential operator' model, which was revived recently under the title δ operator by Goodwin and his co-workers (see Middleton and Goodwin, 1990),

$$x(k) = \frac{B(\delta)}{A(\delta)} u(k-\Delta) \tag{3}$$

where $A(\delta)$ and $B(\delta)$ are polynomials of the following form,

$$A(\delta) = \delta^p + \alpha_1 \delta^{p-1} + \dots + \alpha_p$$

$$B(\delta) = \beta_0 \delta^p + \beta_1 \delta^{p-1} + \dots + \beta_p$$

with the index $p = \max(n,m)$ and the δ operator, for the sampling interval Δt, is the simplest discrete-time differential operation, which is defined as follows in terms of the forward shift operator z,

$$\delta = \frac{z-1}{\Delta t} \; ; \; i.e. \; \delta x(k) = \frac{x(k+1)-x(k)}{\Delta t}$$

Remarks

(1) As $\Delta t \to 0$, the δ operator reduces to the derivative operator $\left(s = \frac{d}{dt}\right)$ in continuous time (i.e. $\delta \to s$).

(2) Given a polynomial of any order n in the z operator, this will be exactly equivalent to some polynomial in δ, also of order n. As a consequence of this, we can easily move between the z and δ operator domains. Also, the δ operator model coefficients (α,β) are related to the forward z operator coefficients (a,b) by simple vector matrix equations (see Chotai et al, 1990).

(3) One attraction of the δ operator model to those who prefer to think in continuous-time terms is that it can be considered as a direct approximation to a continuous-time system. For example, it is easy to see that the unit stability circle in the complex z plane maps to a circle with centre $-1/\Delta t$ and radius $1/\Delta t$ in the complex δ plane; so

that, as $\Delta t \rightarrow 0$, this circular stability region is transformed to the familiar, left-half stability region of the complex s plane (the 'left-half plane'). For very rapidly sampled systems, therefore, the δ operator model can be considered in almost continuous-time terms, with the eigenvalue (pole) positions in the δ plane close to those of the 'equivalent' continuous-time system in the s plane.

A practical example: the ADZ model for transport and dispersion of solutes in stream channels

Classical hydrodynamic analysis associated with dispersion in flowing media, such as the seminal work of G.I. Taylor (1954) on flow in pipes, often provides the starting point for modelling the transport and dispersion of solutes in stream channels. In the purely longitudinal case, this involves a mathematical representation in the form of the following single dimensional, partial differential equation (see e.g. Fischer et al, 1979; Orlob, 1983; Henderson Sellars et al, 1990; Young and Wallis, 1992), usually known as the Fickian Diffusion Equation or the Advection Dispersion Equation (ADE),

$$\frac{\partial x(s,t)}{\partial t} + U\frac{\partial x(s,t)}{\partial s} = D\frac{\partial^2 x(s,t)}{\partial s^2} \qquad (4)$$

where $x(s,t)$ is the concentration of the solute at spatial location s and time t; U is the cross-sectional average longitudinal velocity; and D is the longitudinal dispersion coefficient. The solution to equation (4) for an impulsive ('gulp') input of solute with finite mass M is, with suitable choice for the origins of s and t, given by,

$$x(s,t) = \frac{M}{2A(\pi Dt)^{0.5}} \exp\left\{ -\frac{(s-ut)^2}{4Dt} \right\} \qquad (5)$$

Although this represents a Gaussian distribution in space, when viewed in the time domain at a fixed longitudinal location, it yields a mildly skewed distribution, because the solute cloud continues to evolve as it passes the fixed observation point.

Observations of dispersing solute clouds in rivers have revealed persistent deviations from the behaviour predicted by the classical Fickian theory and the measured concentration profiles rarely, if ever, attain the theoretical Gaussian distribution in space, except at very large distances from the injection point. Partly this is due to fact that it is usual, for logistic reasons, to measure temporal rather than spatial

15

profiles. However, the skewness of the temporal profiles measured in practice is much larger than that predicted by the Fickian theory. The reasons for this are essentially twofold: firstly, observations are rarely made at long enough times after injection for the Gaussian distribution to evolve; and, secondly, real stream channels are non-uniform, and this non-uniformity leads to rather different patterns of mixing than those predicted by the pure Fickian theory.

Derivation of the Aggregated Dead Zone (ADZ) model

The limitations of the ADE have stimulated scientists to either modify the Fickian inspired theory or to develop alternatives. Of these, one approach which has been particularly successful depends on the assumption that 'dead zones' play an important role in the dispersion process. These zones are traditionally associated with side pockets, bed irregularities and other 'roughness' elements in the stream channel; elements which tend to create relatively slower moving regions or even areas of 'backflow'[1]. Solute which is entrained within these areas is then released back into the main flow relatively slowly and at diluted concentrations, due to the 'eddy-like' flow structure which tends to characterize such regions. This induces a distributed time-delay effect and tends to raise significantly the tail of the concentration profile.

Various dead zone analyses (e.g. Valentine and Wood, 1979; Bencala and Walters, 1983) have shown that dead zones can account for some of the observed deviations from the traditional ADE model behaviour. Until recently, however, most of these dead zone models have been based on an extension of the *ADE* model, in which the dead zones are characterized explicitly as a first order differential equation for the dead zone effect, adjoined to the classical partial differential equation (4) of the *ADE*. In the past few years, however, some of the concepts of systems theory have been brought to bear on the problem and this has led to a rather different model formulation: the *Aggregated Dead Zone* (ADZ), where the model is in the form of a lumped parameter, ordinary differential-delay equation for the changes in concentration occurring over a reach of specified length.

This *ADZ* approach to dispersion modelling began in the early 1970s (Beck and Young, 1975, 1976) and came to fruition in the 1980s (Beer and Young, 1983; Young and Wallis, 1986; Wallis et al, 1989). Here, the direct analysis of experimental tracer data, using the recursive estimation procedures described later, is utilized to statistically estimate

---

1. The casual observer who throws a stick into a river may often observe that, near to the bank, the stick can get caught in such a dead zone and may often be seen to move in an opposite direction to the main stream flow.

the parameters in a simple *TF* model, of low or minimal dimension, which is almost always able to explain the measured concentration profile much better than previous conventional models, such as the *ADE*. This *data-based ADZ* approach can be contrasted to more conventional *mechanistic* modelling procedures, such as the *ADE*, where a model structure is derived by physico-chemical reasoning and then utilized as the basis for exercises in model 'fitting' or 'optimization'. Whereas the data-based approach advocated here allows the data themselves to expose the possible model structure within the larger general class of *TF* models, the conventional approach tends to rely on the adequacy of the analyst's preconceptions about the behavioural mechanisms of the system under study.

Although the *ADZ* model is obtained from an objective exercise in statistical identification and parameter estimation, as described below, it can be interpreted best, in physical terms, from its formulation as a continuous-time, first order, differential-delay equation obtained from mass conservation arguments. In previous publications (e.g. Young and Wallis, 1986; Wallis et al, 1989) this equation has been developed by exploiting a 'well mixed' assumption, where the concentration of pollutant in the *ADZ* is assumed to be the same as the concentration at the output of the reach. Here we will use a different but dynamically equivalent approach. If the flow through a reach with total water volume V_a is denoted by Q, then the relationship between the measured concentration $x(t)$ of a conservative pollutant at the output (or downstream) location in the reach to the measured concentration $u(t)$ at the input (or upstream) location, can be obtained from dynamic mass conservation considerations as,

change of mass per unit time	mass flow in per unit time	mass flow out per unit time
$\dfrac{d[V_a y(t)]}{dt}$	$= \quad Q\, u(t-\tau)$	$- \quad Q\, x(t)$

where $y(t)$ is the average concentration of pollutant in the reach and τ is the *advective time-delay* introduced to allow for the translational effects of the unidirectional river flow. In order to convert this to a differential equation relating $u(t)$ to $x(t)$ directly, it is necessary to make some assumption about the relationship between $x(t)$ and $y(t)$. Here, we will make the simplest such assumption, namely that they are linearly related by a factor D_f which we will term the *Dispersive Fraction* for reasons that will become obvious below. In other words,

17

$$y(t) = D_f x(t)$$

so that,

$$\frac{d[V x(t)]}{dt} = Q u(t-\tau) - Q x(t) \qquad (6)$$

where $V = D_f V_a$. Now, if V_a and Q are assumed constant (i.e. steady flow conditions), then equation (6) can be written as

$$T \frac{dx(t)}{dt} = -x(t) + u(t-\tau) \qquad (7)$$

Here $T = V/Q$ secs can be considered as a 'residence time' parameter, in the sense that it provides a measure of the time which the solute, with concentration $x(t)$, resides in the volume $V = D_f V_a$.

The unit impulse response of equation (7) (i.e. temporal response to a unit or 'gulp' input of pollutant applied at $T=0$) is given by,

$$x(t) = 0 \text{ for } t < \tau \; ; \; x(t) = \alpha \exp\{-\alpha(t-\tau)\} \text{ for } t \geq \tau$$

where $\alpha = 1/T$. In other words, the impulsive input is dispersed to yield a concentration profile at the output of the reach which, following the advective time delay of τ time units, jumps to a level $\alpha = Q/V$ and then decreases exponentially with time constant T time units. Thus, the residence time T is associated with the dispersion of the pollutant and provides the main dispersive parameter in the model. We note, however, that T is associated with the volume $V = D_f V_a$ and not the total reach volume V_a; so that if $D_f < 1.0$ then the *effective residence time associated with dispersion processes* is less than the residence time of the reach as a whole. Bearing in mind our hypothesis that it is the dead zones that provide the main mechanism for dispersion, therefore, it is reasonable to associate the volume V with the aggregated effect of all the dead zone behaviour in the reach and refer to it as the *ADZ Volume*. In this manner, we see that the *ADZ* volume V is *not* the total reach volume V_a and will normally be much less than V_a; indeed, as we shall see, the dispersive fraction $D_f = V/V_a$ is normally in the range 0.3 to 0.4 for medium sized natural streams (see later and Young and Wallis, 1986; Wallis et al, 1989).

Clearly, the behaviour of a non-conservative solute could be modelled in a similar manner to the above but the losses occurring within the reach (i.e. both in the *ADZ* and during the advective time delay) would need to be accounted for in some manner, for example by assuming that the loss at any time is proportional to the mass of pollutant in the reach (see Young, 1990); or by introducing chemical/biological reactions.

Finally, it is worth emphasizing that the name 'Aggregated Dead Zone' is rather misleading in its physical connotations. It is a name that was coined originally to draw attention to the relationship between the previous dead zone research and the *ADZ* model. But the *ADZ* is clearly not a 'dead zone' in any real sense of the term; rather it is a lumped approximation to the many 'active mixing zones' that characterize the reach and lead to the dispersion of the pollutant. For this reason, it might be better referred to the aggregated dispersive effect modelled by the residence time *T* as the *Active Mixing Zone (AMZ)* and the associated volume *V* as the *Active Mixing Volume (AMV)*. Indeed, these rather more physically appropriate terms will be used later when we suggest that models of this type have a potentially much wider significance in many different areas of environmental systems modelling.

Transfer Function representation of the ADZ model

The Laplace operator *TF* representation of equation (7) is obtained simply by setting $\dfrac{d}{dt} = s$ and rearranging the equation to yield,

$$x(t) = \frac{\alpha}{s + \alpha}\, u(t-\tau) \tag{8}$$

As we point out later, all discrete-time models can be considered as approximations to their continuous-time equivalents (although this does not imply, of course, that they are any worse at representing the dynamic behaviour of the *real* system being modelled, if they have been obtained by direct estimation from time-series data, as described below). The simplest δ operator approximation in this sense is the following δ-operator *TF* model, which has the advantage of showing the strong similarity between the two types of transfer function,

$$x(k) = \frac{\alpha}{\delta + \alpha}\, u(k-\Delta) \tag{9}$$

Here, for a specified sampling interval Δt, the argument *k* denotes the value of the variable at the *kth* sampling instant (i.e., at a time $k\Delta t$ time units after observation commences); and Δ is the advective time delay in sampling intervals, as defined by the integral value of $\tau/\Delta t$. If $\tau/\Delta t$ is not an exact integral number, then this definition of Δ obviously implies an approximation, the magnitude of which will depend upon the size of the sampling interval. This can be obviated quite straightforwardly but, for simplicity of presentation, we will assume here that τ is an integral number of sampling intervals.

The z^{-1} operator *TF* equivalent of equation (8), for a sampling interval Δt, can be obtained in various ways. If we assume, however, that the input $u(t)$ is approximately constant over the sampling interval, then the discrete-time equation relating the sampled value of the output concentration $x(k)$, at the *kth* sampling instant, to the sampled input concentration $u(k-\Delta)$, Δ sampling intervals previously, is of the form,

$$x(k) + ax(k-1) = b\ u(k-\Delta) \tag{10}$$

where a and b are parameters. In this situation, T can be related to a by the following expression (see e.g. Young, 1984),

$$T = \frac{V}{Q} = -\frac{\Delta t}{\log_e(a)}$$

so that an estimate of the *ADZ* residence time and, given knowledge of the river flow Q, an estimate of the *ADZ* volume V can be derived from a statistical estimate of the parameter a. Finally, the z^{-1} *operator TF* model is obtained simply by introducing the z^{-1} operator into equation (10) and rearranging to yield,

$$x(k) = \frac{b}{1 + a\ z^{-1}}\, u(k-\Delta) \text{ or } x(k) = \frac{b\ z^{-\Delta}}{1 + a\ z^{-1}}\, u(k) \tag{11}$$

which, like the δ operator *TF* model (9), is an approximation to the continuous-time *TF* model (8).

In general, for an arbitrarily selected stretch of river, we cannot assume that a *single ADZ* element such as this will be able to model fully the dispersive properties; it could well be that several elements such as this would be required to model a reasonably long length of river. Depending upon the nature of the river network, this higher order model could be made up of a serial *and/or* parallel connection of first order elements, as shown diagrammatically in Figures 1(a) and (b). It is straightforward to show that such networks of *ADZ* elements can be represented by higher order transfer functions, such as the general model of equation (1) or the more specific models of equations (2) and (3). For example, in the case of (2) the model is characterized by the parameters $a_i, i = 1,2, \ldots, n$ and $b_j, j = 0,1, \ldots, m$. The orders n and m, as well as the time delay Δ, will depend upon the nature of the

20

connection between the elements. In later analysis, it is convenient to use the shorthand notation $[n, m, \Delta]$ to refer to such a model.

The need to specify a general model, such as (2) or (3), which allows for the inter-connections of first order *ADZ* elements, arises from physical considerations. In a single uniform stretch of river, for instance, we might assume that a serial connection of *ADZ* elements, each defined by a first order *TF* such as (11), could describe the dispersive characteristics. However, if we allow for more complicated river networks, or more complex multi-layer structure in each reach of the river channel, then serial and parallel connections are both possible (see Young and Wallis, 1992). In this chapter, we are particularly concerned with the parallel type of connection, such as the simple example in Figure 1(b).

Model identification and parameter estimation

Recursive estimation (see Young, 1984) is probably the most flexible method of modelling *TF* models from time-series data. This approach, which involves sequential updating of the *TF* model parameter, estimates whilst processing the data in a serial manner, has several theoretical and practical advantages over the alternative *en bloc* approaches, such as the better known maximum likelihood methods proposed by Box and Jenkins (1970). In particular, the recursive formulation means that the estimation algorithms can be extended quite easily to handle *nonstationary* models, i.e. models which are characterized by *time-variable parameters* (TVP).

The inherent nonstationarity of many environmental time-series is widely acknowledged and the relevance of such *TVP* methods in environmental systems analysis is obvious. In particular, the algorithms can be used in an *on-line* form as an aid in the *adaptive* management of environmental systems. Typical examples of this currently being developed at Lancaster are an adaptive flood warning system for the Solway River Purification Board; an adaptive flow management model of the River Derwent for the National Rivers Authority; and self-adaptive controllers for greenhouse environmental systems, in association with the AFRC Silsoe Research Institute.

Of the many recursive methods that are now available (see e.g. Ljung and Soderstrom, 1983; Young, 1984; Norton, 1986; Soderstrom and Stoica, 1989) for the estimation of z^{-1} operator transfer function model parameters, only one can also be applied directly to δ and s operator models. This is the *Simplified Refined Instrumental Variable* (SRIV) procedure (Young, 1984, 1985) which exploits special adaptive prefiltering, both to achieve good estimation performance and, in the δ and s *operator* cases, to avoid numerical differentiation. As such, the *SRIV* algorithm provides the most obvious practical vehicle for identification and estimation of *TF* models and it is a central tool in the *microCAPTAIN* time-series analysis computer program package developed at Lancaster and used to generate all the results presented later in this chapter.

The recursive SRIV estimation algorithm

If, for illustrative purposes, we consider the z^{-1} operator case, the adaptive prefiltering which characterizes the *SRIV* algorithm can be justified qualitatively by considering the following stochastic form of equation (2),

$$y(k) = \frac{B(z^{-1})}{A(z^{-1})} u(k-\Delta) + e(k) \tag{12}$$

where $e(k)$ is a zero mean, serially uncorrelated sequence of random variables with variance σ^2. This equation can be written in the following alternative vector form, which is *linear-in-the-parameters* $\{a_i, b_j\}$ of the *TF* model,

$$y(k) = z(k)^T a + \eta(k) \tag{13}$$

where,

$$z(k)^T = [-y(k-1), \ldots, -y(k-n) \; u(k-\Delta), u(k-\Delta-1), \ldots,$$
$$u(k-\Delta-m)] \tag{14}$$
$$a = [a_1 \, a_2 \ldots, a_n \, b_0 \, b_1, \ldots, b_m]^T$$

and $\eta(k)$ is a coloured noise variable defined as follows in relation to the original white noise $e(k)$,

$$\eta(k) = e(k) + a_1 e(k-1) + \ldots + a_n e(k-n)$$

Most estimation problems are posed in a manner such that the variable to be minimized has white noise properties. Thus, a sensible error function is the *response* or *prediction error*, $\hat{e}(k)$,

$$\hat{e}(k) = y(k) - \frac{\hat{B}(z^{-1})}{\hat{A}(z^{-1})} u(k)$$

22

where $\hat{B}(z^{-1})$ and $\hat{A}(z^{-1})$ are estimates of the *TF* polynomials $A(z^{-1})$ and $B(z^{-1})$. Unfortunately, this is nonlinear in the unknown parameters and so the estimation problem cannot be posed directly in simple, linear estimation terms. However, it becomes *linear-in-the-parameters* if we assume prior knowledge of $A(z^{-1})$ in the form of an estimate $\hat{A}(z^{-1})$: then the error equation can be written in the form,

$$\hat{e}(k) = \frac{1}{\hat{A}(z^{-1})} \lfloor \hat{A}(z^{-1})y(k) - \hat{B}(z^{-1})u(k) \rfloor$$

which can be rewritten as,

$$\hat{e}(k) = \hat{A}(z^{-1})y^*(k) - \hat{B}(z^{-1})u^*(k)$$

where,

$$y^*(k) = \frac{1}{\hat{A}(z^{-1})}\, y(k) \; ; \; u^*(k) = \frac{1}{\hat{A}(z^{-1})}\, u(k)$$

are 'prefiltered' variables, obtained by passing $y(k)$ and $u(k)$ through the prefilter $F(z^{-1})$ where,

$$F(z^{-1}) = \frac{1}{\hat{A}(z^{-1})}$$

It will be noted that, if a method for defining $\hat{A}(z^{-1})$ can be defined and the resultant estimation algorithm converges, then this prefilter should have a similar frequency pass-band to that of the system itself: as a result, it performs the intuitively reasonable function of passing those parts of the input and output signals that contain information on the system, whilst filtering off higher frequency 'noise' components that will have a deleterious effect on the parameter estimation.

With the above reasoning in mind, the ordinary recursive *IV* algorithm (e.g. Young, 1984) can be applied *iteratively* to estimate the model parameter vector a, with the variables $y(k)$, $u(k)$ and the instrumental variable $\hat{x}(k)$ replaced, at each iteration, by their adaptively prefiltered equivalents $y^*(k)$, $u^*(k)$ and $\hat{x}^*(k)$, respectively, and with the prefilter parameters based on the parameter estimates obtained at the previous iteration (see Young, 1984, 1985). The main recursive part of the *SRIV* algorithm takes the form,

$$\hat{a}(k) = \hat{a}(k-1) + g(k)\,\{\, y^*(k) - z^*(k)^T\,\hat{a}(k-1)\} \quad \text{(i)}$$

where,

$$\boldsymbol{g}(k) = \boldsymbol{P}(k-1)\hat{\boldsymbol{x}}^*(k) \ [1+\boldsymbol{z}^*(k)^T\boldsymbol{P}(k-1)\hat{\boldsymbol{x}}^*(k)]^{-1} \quad \text{(ii)} \quad (15)$$

and

$$\boldsymbol{P}(k)=\boldsymbol{P}(k-1)+\boldsymbol{g}(k)z^*(k)^T\boldsymbol{P}(k-1) \qquad \text{(iii)}$$

where $P(k)$, which is the inverse of the Instrumental Cross-Product Matrix ($ICPM$; see e.g. Young et al, 1980), i.e.,

$$\boldsymbol{P}(k) = \left[\sum_{i=1}^{i=k} \hat{\boldsymbol{x}}^*(i)^T\boldsymbol{z}^*(i) \right]^{-1} \qquad (16)$$

is also related to the covariance matrix $\boldsymbol{P}^*(k)$ of the estimated parameter vector $\hat{\boldsymbol{a}}(k)$ by the equation,

$$\boldsymbol{P}^*(k) = \sigma^2\boldsymbol{P}(k) \ ; \text{ where } \boldsymbol{P}^*(k) = E\{\tilde{\boldsymbol{a}}(k)\tilde{\boldsymbol{a}}^T(k)\} \qquad (17)$$

and $\tilde{\boldsymbol{a}}$ is the parameter estimation error, defined by,

$$\tilde{\boldsymbol{a}}(k) = \hat{\boldsymbol{a}}(k) - \boldsymbol{a}$$

An estimate $\hat{\sigma}^2$ of the variance σ^2 can be obtained from an additional recursive equation based on the squared values of a suitably normalized recursive innovation sequence (see Young, 1984, p. 100; Young et al, 1991). The standard errors on the parameter estimates are computed from the diagonal elements of $\boldsymbol{P}^*(k)$; so we see that, at every sample in time, the recursive algorithm (15) provides both the currently best $SRIV$ estimate of the model parameter vector $\boldsymbol{a}$ and an estimate of the uncertainty associated with this estimate.

The δ and s operator versions of the $SRIV$ algorithm (15) are quite similar in algorithmic terms but applied to the δ and s operator equivalents of the equation (13). The δ operator algorithm and its use in various automatic control and environmetric applications are described fully in Young et al (1991). The continuous-time versions of the $SRIV$ algorithm, which are closely related to the δ operator versions, are described in Young and Jakeman (1980) and Young (1981); however, they constitute theoretically justified versions of heuristic estimation algorithms developed by the author in the nineteen sixties (Young, 1965, 1970).

SRIV model order identification

The ability to estimate the parameters in a transfer function model represents only one side of the time-series analysis problem. Equally important is the related problem of *Model Order Identification*: namely the identification of the most appropriate values for the orders n and m (or p) of the *TF* polynomials and the size of any accompanying pure time delay which affects the input–output relationship. This is not a simple problem and there is, as yet, no foolproof and entirely objective method for solving it in purely statistical terms. The process of identification can, however, be assisted by the use of well chosen statistical measures which indicate the presence of over-parameterization; i.e. they indicate when the model appears to be characterized by more parameters than are really necessary to explain the observed input–output behaviour.

A reasonably successful identification procedure, which exploits certain useful properties of *IV* estimation (see Young, 1989), is to choose the model which minimizes the following identification statistic,

$$YIC = \log^{e} \left\{ \frac{\sigma^2}{\sigma_y^2} \right\} + \log_e\{NEVN\} \tag{18}$$

where,

σ^2 is the sample variance of the model residuals $e(k)$

σ_y^2 is the sample variance of the measured system output $y(k)$ about its mean value

while *NEVN* is the 'Normalized Error Variance Norm' (Young et al, 1980) defined as,

$$NEVN = \frac{1}{np} \sum_{i=1}^{i=np} \frac{\hat{\sigma}^2 p_{ii}}{\hat{a}_i^2} \tag{19}$$

Here, in relation to the *TF* models (2) and (3), np is the total number of parameters estimated, i.e. $n+m+1$ for model (1) and $2p+1$ for model (2); $\hat{a}_i^2$ is the estimate of the *ith* parameter in the parameter vector **a**, while p_{ii} is the *ith* diagonal element of the $P(N)$ matrix, where N is the sample size (so that $\hat{\sigma}^2 p_{ii}$ is an estimate of the error variance associated with the *ith* parameter estimate after N samples).

This *YIC* statistic is heuristic but can be justified on two grounds: statistical and numerical. From the statistical standpoint, the first term in (18) provides a normalized measure of how well the model explains the data: the smaller the variance of the model residuals in relation to the variance of the measured output, the more negative the first term becomes. Similarly, the second term is a normalized measure of how well the parameter estimates are defined for the *npth* order model: clearly the smaller the relative error variance, the better defined are the parameter estimates in statistical terms, and this is once more reflected in a more negative value for the term. Thus the model which minimizes the YIC should provide a good compromise between model fit and parametric efficiency: as the model order is increased, so the first term tends to decrease; while the second term tends to decrease at first and then to

25

increase quite markedly when the model becomes over-parameterized and the standard errors on its parameter estimates becomes large in relation to their estimated values (in this connection, note that the square root of $\hat{\sigma}^2 p_{ii}/\hat{a}_i^2$ is simply the relative standard error on the *ith* parameter estimate).

From a numerical analysis standpoint, the *NEVN* is one, normalized measure the magnitude of the $P(N)$ matrix which, it will be recalled from equation (16), is the inverse of the instrumental cross-product matrix (*ICPM*). It can be shown (see Young et al, 1980) that, in theory, if the order of the system is greater than that justified by the data, then the *ICPM* will become singular. In practice, complete singularity will rarely occur, and so this problem is revealed by the poor conditioning of the *ICPM*, with the result that the elements of its inverse $P(k)$ can become very large. In this situation, the *NEVN*, whose value is strongly dependent upon $P(k)$, will also increase markedly in value, indicating the poor conditioning of the *ICPM* and associated poor definition (i.e. high variance) of the *TF* model parameters.

In practical applications, of course, the minimization of *YIC* will not, in itself, guarantee that the 'best' model has been identified: it is, after all, only one possible, heuristic statistic which is naturally dependent upon the quality and size of the time-series data set. Consequently, inadequate or very noisy data can lead to a YIC identified model structure which may not be acceptable for some good physical reasons. In other words, the *YIC* should be used, together with other *physically motivated* considerations, such as the nature of the model response characteristics and any *a priori* information on the system dynamics, as a *guide* to the selection of the most appropriate model order. For example, the *YIC* can tend to under-identify the model order when the input signal is limited for some reason; e.g. in tracer experiments where it is restricted to an impulsive 'gulp' or single step excitation. In such circumstances, where the input is not *persistently exciting* (see e.g. Young, 1984), it may well be that higher order (and usually better fitting) models, with somewhat larger *YIC* values, will have certain practical advantages. It would then be advisable to select one of these higher order models, particularly if *a priori* considerations suggest that its superior fitting ability outweighs the possibility of over-parameterization. Practical examples of such discrimination are discussed in the next example section of the chapter.

Practical examples

In this section, we consider five practical examples from the area of hydrology and water quality where the methodological approach to time-series modelling described in previous sections reveals the likely existence of parallel pathways between the measured input and output of the system. The first example explores further the *ADZ* model by considering its application to the analysis of data from dye-tracer experiments carried out in the River Conder, near Lancaster in North West England. The example is dealt with fairly briefly since the results have been discussed fully in the cited references (i.e. Young and Wallis, 1986; Wallis et al, 1989). The second two examples are concerned with the modelling of rainfall-flow processes in two catchments, one in Wales and the other in France. The final two examples consider the modelling of soil-water processes based on breakthrough curve data obtained from tracer experiments on saturated soil columns.

Example 1: Systems models of pollutant dispersion in rivers

Let us consider first the *ADZ* modelling results obtained from the analysis of dye-tracer data from an experiment carried out on a short reach of the River Conder, a medium sized stream in North West England. Further details of the data are to be found in Young and Wallis (1986) and Wallis et al. (1989). The reach was 5 metres wide and 116 metres in length, characterized by one major pool-riffle structure and a fine cobble bed, and samples were taken every 15 seconds. A number of different order *TF* models were estimated from these data using the *SRIV* method of recursive estimation in the *MicroCAPTAIN* microcomputer program (see Young and Benner, 1991). Table I compares the results obtained from this exercise and shows that the best identified model in terms of the *YIC* identification statistic is the first order (i.e. $[1,0,12]$ or $n=1$; $m=0$; $\Delta=12$) model, which can be interpreted in physical terms as a

Table I *SRIV* identification results

Den. Order n	Num. Order m	Advective Delay δ (sampling ints.)	R_T^2	YIC Ordering
1	0	12	0.9933	1
2	0	11	0.9923	4
2	2	12	0.9978	2
3	0	11	0.9971	5
3	1	11	0.9981	

single *ADZ* element with a residence time of 65 seconds and an advective time delay of 12 sampling intervals (i.e. 180 seconds).

An alternative model is the second order model [2,2,12], which can be interpreted as two *ADZ* elements with residence times of 46 and 182 seconds respectively. This second order model has a somewhat larger YIC value but it fits the data rather better, with a Coefficient of Determination $R_T^2 = 0.9978$ compared with 0.9933 for the first order model (i.e. 99.78% of data explained), particularly in the important tail of the concentration profile.

Clearly, from a purely objective *YIC* standpoint, we should accept the first order model as the most appropriate representation of the data, with the better fit provided by the higher order model not fully justified in statistical terms by its increased parametrization. However, we see that the second order model is quite well identified, being the second choice in *YIC* terms, and the improvement in the explanation of the elevated tail seems to be visually, and as we discuss below, physically significant.

This improved modelling of the tail portion of the concentration profile is particularly important in the present context since it is a direct product of the model's parallel structure. If we assume that the second order model is the best candidate, then it can be decomposed, quite unambiguously, into two parallel *ADZ* elements which have considerably different residence times and are arranged in a *parallel* structure, as shown in Figure 1(b). A rather appealing, albeit speculative, interpretation of such a model structure, therefore, is a two layer flow pattern, with the longer residence time *ADZ* representing a slower moving layer close to (or within) the bed, and the shorter one representing the faster moving layer above this. This seems quite a reasonable conjecture on the basis of the River Conder, where the fine cobble bed could create the right conditions for such multi-layer behaviour.

This example emphasizes our earlier comments that, while objective statistical criteria such as the *YIC* can be a great help in ensuring that the model is not over-parameterized, they should not provide the *only* criteria on which the model is judged. If the model has a reasonable physical interpretation, as here, then physical considerations may well be used to over-ride the purely statistical considerations, and add to the model's credibility. This seems quite reasonable from a Bayesian statistical standpoint,[2] for example, which would suggest that prior knowledge of this type should be allowed to influence the modelling.

2. It is interesting to note that the recursive approach to time-series analysis used here can be considered directly in Bayesian terms; indeed, the recursive update at each sample from the *a priori* to *a posteriori* estimate of the parameter vector is the very embodiment of Bayesian estimation (Young, 1984).

But we should not allow this influence to be too great without careful thought: in this example, for instance, the reason for the ambiguity between first and second order models arises partly from the poor excitation provided by the impulsive input and so, before reaching a final decision on which model is most appropriate, we might think of conducting other experiments to clarify the situation (see Jakeman and Young, 1980). Indeed, in this case, numerous experiments were carried out on the River Conder over a wide range of discharge conditions (see below) and the parallel structure was indicated most of the time. It should be emphasized, however, that either of the two best identified models (first or second order) would perform quite well in simulation or predictive applications, since they both explain over 99% of the experimental data (much better than an *ADE* model) and are reasonably efficient in their parametrization.

Wider implications of the ADZ model

One very interesting aspect of the *ADZ* modelling studies is the manner in which the model characteristics change with river discharge. In particular, while both the *ADZ* volume V and the total volume of water V_a in the reach vary over a wide range, as might be expected, the fractional volume of the reach which appears attributable to *ADZ* effects, the dispersive fraction D_f introduced in the previous section on the *ADZ* model, is approximately constant at a value of 0.37 over the whole range of flows investigated. In other words, only 37% of the river volume appears primarily responsible for introducing the pollutant dispersion effects, with the remainder functioning in a largely advective role.

It is easy to show (see e.g. Wallis et al, 1989) that, within a more conventional hydrological context, D_f is also equivalent to the ratio of the *ADZ* residence time to the travel time (where travel time is simply the sum of the *ADZ* residence time and the pure advective time delay). Also, the results can be linked with the earlier work of Day (1974) and Day and Wood (1976): they normalized the results of the impulsive tracer experiments by introducing adimensional concentration and time measures x_a and t_a, where,

$$x_a = \frac{x}{x_m} \text{ and } t_a = \frac{t - t_1}{t_2 - t_1}$$

Here x_m is the maximum concentration of the measured impulse response; and t_1, t_2 are, respectively, the times taken to reach half the maximum concentrations on the rising and falling limbs of the impulse response. Day and Wood applied this normalization to over 700 tracer profiles, obtained from different reaches and under different flow conditions; and they found that, within the limits of experimental accuracy, all the normalized curves conformed to a common pattern, with the characteristic shape we have discussed in earlier sections: i.e. with a fairly rapid rise to the maximum concentration (normalized at unity) and a more slowly decaying tail portion. Ribeiro da Costa (1992) has obtained similar results for the River Ave in Northern Portugal.

In effect, the Day and Wood analysis can be considered as one particular graphical interpretation of tracer experimental data that reflects certain aspects of *ADZ* modelling. In particular, the efficacy of the *ADZ model* is indicative of an underlying similarity in the profile shape and this is inherently exploited by Day and Wood in their graphical normalization. However, their normalization is only concerned with the dispersive part of the response and does not include the advective component of the response. As a result, it is not surprising that our experimental results suggest that D_f tends, over a wide range of different stream channels, to be river dependent; with values ranging from around 0.4 for fast, turbulent rivers, to 0.15 for smooth, man-made channels.

These *ADZ* modelling results show the potential value of the time-series approach to modelling. By analysing the experimental tracer data in a careful and largely objective, statistical manner, it is not only possible to model the data efficiently, but also expose an important flow-invariant parameter; namely the dispersive fraction. It is always satisfying if such an invariant characteristic can be discovered in scientific investigations and it is already helping to stimulate new theoretical (see e.g. Smith, 1987) and practical developments on dispersion modelling. In practical terms, studies carried out with North West Water, Yorkshire Water and the Water Research Centre have shown the potential value of the model in incident prediction and the evaluation of storm water overflows.

Finally, it should be noted that the kind of *ADZ* modelling discussed here can be considered as a method of calibrating the river in terms of its transport and dispersive properties. As a result, the model can form a useful basis for the development of more complex water quality models for interacting, non-conservative pollutants or biological variables (see e.g. Young, 1990; Ribeiro da Costa, 1992).

Example 2: Rainfall-flow processes – a UK catchment

Figures 2(a) and (b) are plots of 400 hourly rainfall and flow measurements in a Welsh upland catchment (Jakeman et al, 1990). The modelling of such rainfall flow processes is, of course, a major problem in flow and flood forecasting applications (see e.g. Weyman, 1975; Kraijenhoff and Moll, 1986). Visual inspection of the data in Figure 2 quickly confirms the well known fact that the physical process involved is nonlinear, since 'antecedent' rainfall conditions clearly affect the subsequent flow behaviour. The system's contribution to the modelling of such processes is linked with the development of discrete-time *TF* modelling procedures such as those discussed in previous sections. But since such models are essentially linear, it proves necessary to introduce the soil moisture nonlinearity in some manner. Young (1975), White-head et al (1979) and Jakeman et al (1990), for example, proposed the following nonlinear model for effective rainfall *u(k)* at a *kth* discrete instant of time,

$$
\begin{aligned}
&u(k)S(k)r^*(k) &&\text{(i)} \\
&S(k) = S(k-1) + [1/T_s]\{r^*(k) - S(k-1)\} &&\text{(ii)} \qquad (20) \\
&r^*(k) = \gamma(T_m - T_i)r(k) &&\text{(iii)}
\end{aligned}
$$

where *r(k)* is the rainfall; *S(k)* is a measure of soil moisture; $T_m - T_i$ is the difference between the overall maximum temperature T_m and the monthly mean temperature T_i for the month in which the *kth* observation is being taken; *r*(k)* is a transformed measure of rainfall which allows for temperature induced, evapo-transpiration effects; and γ is a parameter associated with this evapo-transpiration transformation.

The overall effect of the nonlinear transformation (20) is that equation (iii) yields a rainfall measure *r*(k)* which is adjusted to allow for the predominantly seasonal evapo-transpiration losses; equation (ii), which is simply the low pass filtering (exponential smoothing) of *r*(k)* with time constant T_s, provides a measure of soil moisture with T_s representing a 'time constant' associated with the wetting-drying processes;[3] and the multiplicative nonlinear transform in (i) yields an effective rainfall measure which is now compensated for both of the major physical processes thought to be involved in the rainfall-flow dynamic system.

In practice, over shorter periods of time such as shown in Figure 2, the evapo-transpirative effect is small, so that *r*(k)* in (ii) can simply be replaced by *r(k)*; in which case the nonlinearity is assumed to depend entirely on the soil moisture changes and, in particular, on the time constant T_s. Recognizing that river flow *y(k)* is itself naturally low pass filtered rainfall, the author (Young and Beven, 1991) has proposed an

---

3. Exactly equivalent to an infinite length, exponentially weighted antecedent precipitation index.

alternative and simpler 'bilinear' model in which $r^*(k)$ in (i) is replaced by the flow measure $y(k-d)$, where d is normally zero, so that there are no longer any unknown parameters in the nonlinear part of the model. In effect, this is equivalent to assuming that the natural time constant associated with the rainfall-flow process will also be reflected in the soil moisture dynamics.

The resultant 'effective rainfall' in this case, for $d = 0$, is shown in Figure 3, which can be compared with Figure 2(b): we see that the main effect of the nonlinear transform is to adjust the heights of the rainfall peaks to reflect the prevailing flow at the same sampling instant; so amplifying peaks where flow is high and attenuating those where flow is low. Having accounted for most of the nonlinear aspects of the rainfall-flow process, it is now possible to model the remaining *linear dynamic* relationship between the effective rainfall in Figure 3 and the flow using *TF* model identification and estimation.

In this first rainfall-flow example, we will consider both z^{-1} and δ operator *TF* modelling results. In the former case, the resulting estimated *TF* model takes the form,

$$y(k) = \frac{19.9363 - 19.0087 \, z^{-1}}{1 - 1.7108 \, z^{-1} + 0.7149 \, z^{-2}} \, u(k) + \xi(k) \qquad (21)$$

where $\xi(k)$ represents noise on the data. This model explains the data quite well with a Coefficient of Determination, $R_T^2 = 0.944$. As in the *ADZ* example above, this model can be decomposed unambiguously by partial fraction expansion into the parallel structure shown in Figure 4, where we see that the model suggests that some 28% of the effective rainfall on the catchment reaches the river rapidly (time constant or residence time $T = 3.1 \, hrs$), presumably as surface or 'quick-flow'; while 72% reaches the river much more slowly ($T = 66.2 \, hrs$), probably via the groundwater, as 'slow-flow' or baseflow (although the exact interpretation of these mechanisms is open to argument; see Young and Beven, 1991).

These z^{-1} operator model results are confirmed by the alternative δ operator model analysis. The resulting *TF* model in this case is again identified as second order and takes the form,

$$y(k) = \frac{19.5821 \, \delta + 0.6927}{\delta^2 + 0.2626 \, \delta + 0.00278} \, u(k) + \xi(k) \qquad (22)$$

which yields an $R_T^2 = 0.943$. A comparison of the model output and the measured flow is given in Figure 5, which is very similar to the result obtained with the z^{-1} operator model, as indicated by the virtually identical R_T^2 values. Again this is a parallel model, as shown in the block diagram Figure 6, which can be compared directly with Figure 4.

It is interesting to note that this method of modelling rainfall-flow processes seems to allow the modeller to objectively estimate the 'baseflow' element of the river flow: this is simply obtained as the output $x_2(k)$ of the lower TF in Figures 4 and 6. This can then supplement other methods of baseflow estimation used in conventional hydrological analysis. However, a word of caution is necessary in this regard: note that there are apparently significant differences in the dynamic nature of the two pathways revealed by Figures 4 and 6: although the estimated partitioning of the flows is identical, the gains and time constants in the pathways are quite different, this despite the fact that both models fit the data almost identically! So which model should we use for evaluating the partitioning and baseflow component?

The above results suggest strongly that, while both z^{-1} and δ operator models are able to explain the data well, the parallel decomposition is not too well defined, so that the 'baseflow' element will differ somewhat depending on which model is utilized in the computation. In systems terms, we conclude that this aspect of the system behaviour is *not well identified* from the data. This poor identifiability is confirmed if we carry out a Monte-Carlo sensitivity study of the z^{-1} operator TF model. In this analysis, the decomposition of the model (21) is carried out repeatedly, each time with the model parameters selected randomly from a normal probability distribution defined by the $SRIV$ estimation results; i.e. with mean and covariance defined by the estimated parameters and covariance matrix $(\hat{a}(N), P^*(N))$. The estimated percentage partitioning of flow in the two pathways is then analysed for each random realization and the frequency histograms of the results based on 1000 repetitions are shown in Figure 7. Also shown as vertical lines are the specific partitioning results obtained from the nominal z^{-1} and δ operator TF models obtained here (28%:72% partition), as well as the results (40%:60% partition) obtained by Jakeman et al (1990) using a nonlinear effective rainfall definition similar to equation (20).

It is clear that, while there is very good evidence from the data of a parallel pathway process, the exact nature of this process is not well defined. It is also clear that *any*[4] model identified from these same data will suffer from similar identifiability problems, unless there is *additional*

---

4. i.e. the poor identifiability is endemic in the data: the present model is no worse for the fact that we have exposed the uncertainty on the flow partitioning; indeed, other more complex 'physically-based' simulation models *based on the same data set* may be even more uncertain in this regard because of over-parameterization but the modeller may not be aware of this.

information that can remove the uncertainty about the magnitude of the flow partitioning. In other words, while the data set in Figure 2 are sufficient to confirm our intuition about the presence of parallel flow processes, they do not, in themselves, appear to contain sufficient information to provide accurate estimates of how much flow is passing through the two major pathways or, indeed, on the exact physical nature of the pathways (see the discussion in Young and Beven, 1991).

The poor identifiability encountered in this example is almost certainly a by-product of two major factors: uncertainty about the best way to quantify the nonlinear aspects of the model; and the limited impulsive nature of the rainfall inputs which probably provide insufficient information over the observation period on the 'slow-flow' process (note that the 400 hours of data only span between four and seven of the 'slow flow' model time constants, depending upon which model is considered). This latter problem is accentuated in the next example, where the data are even more limited in this regard.

Example 3: Rainfall-flow processes – a French catchment

Figure 8 shows 5 estimated impulse responses (unit hydrographs) obtained by *SRIV* analysis of rainfall-flow data from a river catchment in France and kindly made available to the author by Professor Charles Obled of Grenoble University. Unlike the example in the previous section, the data here are not in the form of a continuous record, but consist of five fairly short sequences of rainfall-flow records covering individual storm events. As a result, it is not possible to analyse them in exactly the same manner as the first example, and each limited set of rainfall-flow data is considered individually using the *SRIV* approach. Although Figure 8 shows that the impulse responses of each model vary quite a lot, it is interesting to note that most exhibit a small secondary peak or 'shoulder' which could be indicative of a parallel flow mechanism.

In an attempt to overcome some of the limitations of the data set as regards parallel process identification, *SRIV* identification and estimation was also applied to the *average* estimated impulse response. This yields the following *TF* model,

$$y(k) = \frac{0.257 - 0.214 \, z^{-1} - 0.0095 \, z^{-2}}{1 - 2.364 \, z^{-1} + 1.841 \, z^{-2} - 0.474 \, z^{-3}} u(k-2) + \xi(k) \quad (23)$$

which fits the average response well, as shown in Figure 9, with $R_T^2 = 0.995$. The poles of the denominator in this model are at *0.959*, *0.702±0.030j* in the complex z plane which, within the uncertainty on the parameter estimates, can be considered as three *real* roots at *0.959, 0.702, 0.702.*

This approximate model indicates a slow mode with time constant *28.39* samples, and two quick modes in series with each other, both with time constant *2.83* samples (or total effective time constant of the two equal real roots = *5.66* samples). As in previous examples, this model can be decomposed unambiguously into the following parallel form,

$$y(k) = \frac{0.3272}{1 - 0.959 \; z^{-1}} u(k) + \frac{-0.0702 + 0.1718 \; z^{-1}}{(1 - 0.702 \; z^{-1})^2} u(k) + \xi(k) \quad (23a)$$

and its impulse response is virtually identical to that of model (23), as shown in Figure 9. It is interesting to note that the -0.0702 parameter in the numerator of the second *TF* on the right hand side of (23a) means that it is mildly non-minimum phase in character: this can be considered as a first order Padé approximation to a one period, pure time delay and suggests that the second order mode is characterized by such a delay, which is physically quite reasonable.

It must be emphasized that the results presented in this example are of only a preliminary nature, based on an initial short analysis of the data. Further evaluation of the data is continuing in association with Professor Obled, whose group is well known for its own novel approach to rainfall-flow modelling and forecasting which involves differencing data to remove the baseflow effects.

Example 4: Solute transport in soils I

This example concerns the transport of solutes through saturated soils and is based on the paper by Beven and Young (1988). The manner in which solutes are transported through soils is clearly an important factor in determining how dissolved pollutants will be transported and dispersed in the natural environment. The conventional approach to modelling the transport and dispersion of solutes in soils has been by recourse to the *ADE* model (4), which is usually termed the Convective-Dispersion Equation (*CDE*) when applied to soil-water problems. This model has often proven useful in describing the observed dispersion of solutes with adequate accuracy, at least for repacked columns of soil in the laboratory. As in the case of dispersion in rivers and open channels, however, there has been growing awareness that solute transport in more realistic conditions, for example through soils in the field and in undisturbed soil columns, does not often conform to the predictions of *CDE* theory. In particular, and again in sympathy with the situation of streamflow dispersion, the graph of the concentration of the solute in the water leaving an undisturbed soil column, or the *breakthrough curve* as it is termed in the literature, tends to show more rapid initial rise and a longer tail than expected, following the application of an impulsive input of solute at the top of the column (e.g. see Elrick and French, 1966; Smettem, 1984; and many others).

The situation encountered here is very similar to that in the case of riverine pollutant dispersion and it is not surprising that the system's contribution to the subject is closely related to that used in rivers, namely the *ADZ* model described previously in this chapter. In applying *ADZ* concepts to soils, it is assumed that there may be inefficient mixing between the faster flow pathways in larger structural pores and slower flows in the matrix, but that an aggregate effective mixing volume again may be able to model the resulting dispersion. However, it is more likely in the soil that there will be part of the soil water storage that takes no part in the transport process (except by slow molecular diffusion) and so acts as a true dead zone. Consequently, in order to avoid confusion with more traditional concepts, we have used the name *Aggregated Mixing Zone* (*AMZ*) to describe the model (Beven and Young, 1988).

As in the riverine *ADZ* case, the main limitation of this *AMZ* model lies in the fact that it is essentially a linear, small perturbation model similar to earlier transfer function models introduced by Jury and his co-workers (e.g. Jury, 1982; Jury and Sposito, 1986). The *AMZ* model, however, has the advantage that it is obtained using a data-based approach based on recursive estimation which, as we have seen, allows for objective identification of the model order and the ability to estimate any changes in the parameters that might be expected as the flow régime changes.

Most of the *AMZ* models obtained by Beven and Young (1988) are of second or third order and their bi-modal impulse response shape again suggests the presence of parallel flow pathways. In the case of Figure 4 in the Beven and Young paper, for example, the third order model can be unambiguously decomposed into the following form,

$$y(k) = \frac{b_1 z^{-1}}{1 + a_1 z^{-1}} u(k-2) + \frac{b_2 z^{-1} + b_3 z^{-2}}{1 + a_2 z^{-1} + a_3 z^{-2}} u(k-2) + \xi(k) \quad (24)$$

where $y(k)$ is the concentration of solute in the water leaving the soil column and $u(k)$ is the input concentration (in this experiment, an impulse input). In other words, the model suggests a parallel connection of first and second order sub-systems: the relatively fast, first order system (the first term on the right hand side of equation (24)) has a mean *AMZ* residence time (time constant) of 599 secs.; while the slower, second order system is composed approximately of two first order systems *in series*, each with *AMZ* residence times of 943 secs.

The estimated residence times in this model imply *AMZ* volumes of 53.9 cm^3 and 94.4 cm^3 respectively, and the total effective mixing volume of all *AMZ* elements is found to be 21% of the soil sample, which was close to saturation in the experiment. An estimate of the porosity of the sample was not available in this experiment, but it is expected that this effective mixing volume is much less than the water filled pore space in the sample, suggesting significant by-passing behaviour in the solute transport process.

Example 5: Solute transport in soils II

In this final example, we again consider the modelling of solute transport in an experimental soil column based on data collected some years ago. However, since the full results have not been published previously and represent a more difficult problem of parallel flow model decomposition, we discuss them in somewhat more detail. The breakthrough curve data are shown in Figure 10, and Table II shows the identification results obtained by analysis of these data using recursive least squares, instrumental variable and *SRIV* methods. Figure 11 compares the best model fits obtained with the three different methods; and Table III presents the detailed estimation results for some of the more interesting examples listed in Table II.

Table II and Figure 11 demonstrate the power of *SRIV* estimation, which is able to explain the data almost perfectly: in contrast, the least squares (*LS*; effectively linear regression) model yields a very poor fit; and the ordinary *IV* model yields an improved result, but not nearly so good as the *SRIV* model. However, Table III reveals that the *LS*, *IV* and the [3, 3, 0] *SRIV* models are all characterized by significantly complex poles (eigenvalues). Since there is no obvious physical mechanism for the oscillatory modes associated with these complex poles, this calls into question their efficacy. However the [2, 5, 0] *SRIV* model (not listed in Table II) also fits the data almost perfectly and, while it nominally has complex poles, the imaginary part is small so that the model is well approximated by two *real* poles, both at 0.87.

What do these estimation results reveal? From the data, we see that the process is almost certainly exhibiting parallel flow behaviour, with the secondary peak exceptionally significant and strongly redolent of some tracer passing through a slower pathway. But since the impulse response is so complex in this regard, the estimation algorithms are clearly having difficulty in quantifying the nature of the parallel partitioning: it is, yet again, almost certainly an example of poor identifiability caused by the paucity of data (there are, after all, only 40 data points and the input is a single sample impulse).

We can, however, proceed further with the analysis, albeit in a manner which exploits the subjective view that parallel activity is present. To do this, we first observe that the initial peak is probably the result of a first order mode with zero pure time delay; while the more rounded second peak is primarily the result of a higher order, slower mode with 2 or 3 sample pure delay, which becomes prominent as the effects of the first order mode decay. Consequently, it makes sense to attempt to exploit the recursive algorithm's ability to handle missing data and model the first order mode on the basis of only the first few data samples, with samples 13 to 40 completely removed, but retaining samples 40 to 50, since by then the effects of the second order mode will have decayed away. The results of this estimation are given in Table IV, together with those obtained by fitting a [3, 1, 2] model to the *original data* but *with the response of the first order estimated mode removed by subtraction* prior to estimation. The poles of this third order model are at *0.875* and *0.821±0.19j* in the complex z plane and, since the imaginary parts are fairly small, we see that, within the uncertainty bounds on the estimates, the model is closely approximated by three *real* poles, as required from physical considerations.

The final model obtained by this two step decomposition approach takes the following form,

$$y(k) = \frac{1.102}{1 - 0.668 \, z^{-1}} \, u(k) +$$

$$\frac{0.083z^{-2}}{1 - 2.516 \, z^{-1} + 2.147 \, z^{-2} - 0.662 \, z^{-3}} \, u(k) + \xi(k)$$

and the response of the two parts of the model are compared with the data in Figure 12. The composite response, as obtained by summing these two individual responses, fits the data very well (similar to the initial *SRIV* model), confirming the effectiveness of the decomposition. It is clear from these results that the data could well have been generated by the two parallel pathway model shown in Figure 13, with the fast pathway dominated by a first order *AMZ* process having a residence time of 2.5 samples; and the slower pathway consisting of three, similar, first order *AMZ* processes, each with a residence time of 7.2 samples (i.e. *21.6* samples in all). This model yields a flow partitioning of 26.5% down the fast pathway and 73.5% down the slower pathway, again with some degree of uncertainty.

Although the conjecture of parallel flow with the partitioning in

Table II Soil core identification results

(a) LS algorithm			(b) IV algorithm			(c) SRIV algorithm		
Model	R_T^2	YIC	Model	R_T^2	YIC	Model	R_T^2	YIC
3,3,0	0.788	−1.847	3,3,0	0.978	−6.699	3,3,0	0.999	18.486
4,4,0	0.995	3.08	4,4,0	0.994	−4.690	4,4,0	0.999	−11.24
4,3,0	0.830	−1.077	4,3,0	0.985	−5.883	4,3,0	0.998	− 6.119
3,4,0	0.883	−1.444	3,4,0	0.980	−5.238	3,4,0	0.998	− 6.118

Figure 13 has not been proven by the analysis outlined here, it seems a reasonable supposition if we consider the physical nature of the system. At the very least, it seems a satisfactory result with such a limited data set. Of course, the best approach at this stage would be to carry out more experiments, if possible with better input excitation, in order to confirm the results obtained here.

The Active Mixing Volume (AMV)

The examples in the previous sections demonstrate the wide potential of the modelling procedures discussed in this chapter. However, recent research suggests that the concept of an *Active Mixing Volume*, as discussed earlier in relation to the *ADZ* model, has even wider theoretical and practical significance. In particular, if the idea of an active mixing volume applies in the context of dynamic mass conservation in flowing media, might it not be applicable in the characterization of other important flow phenomena, such as heat flow modelled via the formulation of energy balance equations (see Young and Lees, 1992)? Figure 14 is a diagrammatic representation of the *AMV* concept in these more general terms. Here, it is assumed that the flow process, whatever its precise form, will normally be rather inefficient, in the sense that the *AMV* will, in practical situations, be less than the actual volume V_a. Because it provides the main dispersive mechanism, however, it is this *AMV* which will dictate the observed dynamic behaviour of the system; be it the changes in the concentrations of pollutants in a river; the variations in temperature in soil; or any other flow process of this general type.

The implications of such an hypothesis are profound. Indeed, if it does prove as widely applicable as we suggest here, it could change the manner in which environmental systems are modelled via the formula-

Table III Selected soil core estimation results
(figures in parentheses are standard errors)

LS Estimates
[3,3,0]

$\hat{a}_1 = -2.051(.825)$; $\hat{a}_2 = 1.444(1.248)$; $\hat{a}_3 = -0.379(.487)$: roots (den)
0.953; 0.55+0.31j
$\hat{b}_0 = 1.114(.140)$; $\hat{b}_1 = -1.585(.931)$; $\hat{b}_2 = 0.689(.856)$: roots (num) 0.953;
0.55+0.31j

IV Estimates
[3,3,0]

$\hat{a}_1 = -2.375(.326)$; $\hat{a}_2 = 1.942(.486)$; $\hat{a}_3 = -0.554(.181)$: roots (den) 0.909;
0.78+0.27j
$\hat{b}_0 = 1.114(0.046)$; $\hat{b}_1 = -1.946(.366)$; $\hat{b}_2 = 1.017(.327)$: roots (num) 0.953;
0.55+0.31j

SRIV Estimates
[3,3,0]

$\hat{a}_1 = -2.3975(.005)$; $\hat{a}_2 = 1.9464(.009)$; $\hat{a}_3 = -0.5358(.004)$: roots (den)
0.834; 0.78+0.18j
$\hat{b}_0 = 1.115(.005)$; $\hat{b}_1 = -1.976(.01)$; $\hat{b}_2 = 1.0224(.006)$: roots (num) 0.953;
0.55+0.31j

[2,5,0]

$\hat{a}_1 = -1.7453(.013)$; $\hat{a}_2 = 0.77163(.012)$: roots (den) 0.8727 + 0.1j
$\hat{b}_0 = 1.1133(.027)$; $\hat{b}_1 = -1.2425(.055)$;
$\hat{b}_2 = 0.1538(.057)$; $\hat{b}_3 = 0.071(.055)$; $\hat{b}_4 = 0.2213(.034)$

Note: almost two real roots at 0.87: i.e. $(1-0.87z^{-1})(1-0.87z^{-1}) =$
$1 - 1.74z^{-1} + 0.76z^{-2}$

---

tion of dynamic conservation equations. No longer could it be assumed that the volumes which are conventionally associated with the medium under consideration, and which appear in the conservation equations, are the actual, measured volumes and are, therefore, known *a priori*: rather they would simply become other, unknown parameters in the model. Such *AMV* parameters would then be seen as important properties of the medium; properties which, within the natural environment, might well vary considerably, depending upon the nature of the medium at different geographical locations. And, as such, they

Table IV Decomposition estimation results
(figures in parentheses are standard errors)

SRIV Estimates
[1,1,0]
(with missing data from sample 13 to 40)
$\hat{a}_1 = -0.6678(.005)$
$\hat{b}_0 = 1.102(.007)$

[3,1,2]
(with first order [1,1,0] process effect removed)

$\hat{a}_1 = -2.5163(.011); \hat{a}_2 = 2.147(.020); \hat{a}_3 = -0.622(.009)$
$\hat{b}_0 = 0.0829(.002)$

roots (den) 0.875; 0.821 + .19j: note almost three equal real roots

would need to be determined empirically, using either a procedure similar to that discussed in this chapter, or some other, alternative estimation techniques.

Conclusions

The chapter has outlined one particular approach to data-based mechanistic modelling that has proven very useful in many environmental applications at Lancaster over the last few years. By means of several practical examples in the areas of hydrology and water quality, it has also demonstrated how the model builder can exploit this approach to rapidly extract important dynamic information from limited sets of time-series data. In particular, these examples have demonstrated how the existence and nature of important parallel processes in hydrological and water quality systems can be quickly investigated once time-series data become available. Such analysis can be useful in its own right, for applications such as real-time flood forecasting and warning; or it can be used as a prelude to the construction of more complex mechanistic simulation models.

The chapter also introduces the hydrological audience to two new ideas: delta (δ) operator models, which are believed to have good potential for application in hydrological systems analysis when modelling dynamic processes from rapidly sampled or continuous-time data; and the Active Mixing Volume (AMV), a general concept which we feel has important theoretical and practical implications on the formulation of conservation equations for flow processes, not only in hydrology but in all areas of environmental systems analysis.

Acknowledgements

The author is grateful to Dr A.J. Jakeman of the Centre for Resource and Environmental Studies at the Australian National University in Canberra for the data used in Example 2; and to Professor Charles Obled of the Civil Engineering Department at the University of Grenoble, France, for the data used in Example 3 and for many interesting discussions during his recent sabbatical visit to Lancaster. The research described in this chapter has been funded in part through Natural Environment Research Council (NERC) and Science and Engineering Research Council (SERC) grants and the author is very grateful for this support. He and his colleague Dr Keith Beven are also grateful for generous support from other sources, including North West Water PLC, Yorkshire Water PLC, the Water Research Centre, and the Solway River Purification Board.

References

Beck, M., and Young, P.C. (1975) A dynamic model for DO-BOD relationships in a non-tidal stream, *Water Research*, **9**, 769–776.

Beck, M.B., and Young, P.C. (1976) Systematic identification of DO-BOD model structure, *Jnl. Env. Eng. Div., American Soc. Civil Eng.*, **102 (EE5)**, 902–927.

Beer, T., and Young, P.C. (1983) Longitudinal dispersion in natural streams, *Jnl. Env. Eng. Div., American Soc. Civil Eng.*, **109**, 1049–10.

Beven, K.J., and Young, P.C. (1988) An aggregated mixing zone model of solute transport through porous media, *Jnl. of Contaminant Hydrology*, **3**, 129–143.

Box, G.E.P., and Jenkins, G.M. (1970) *Time Series Analysis, Forecasting and Control*, Holden-Day, San Francisco.

Chotai, A., Young, P.C., and Tych, W. (1990) A non-minimum state space approach to true digital control based on the backward shift and delta operator models, in: M.H. Hamza (ed.), *Proc. IASTED Conference 1990*, Acta Press, Calgary, 1–4.

Day, T.J. (1974) Longitudinal dispersion in natural channels, *Water Resources Research*, **11**, 909–918.

Day, T.J., and Wood, I.R. (1976) Similarity of the mean motion of fluid particles dispersing in a natural channel, *Water Resources Research*, **12**, 655–666.

Elrick, D.E., and French, L.K. (1966) Miscible displacement patterns on disturbed and undisturbed soil cores, *Soil Sci. Soc. Amer. Jnl.*, **30**, 153–156.

Fischer, H.B., List, E.J., Koh, R.C., Imberger, J., Brooks, N.H. (1979) *Mixing in Inland and Coastal Waters*, Academic Press, New York.

Gould, R.P., Minchin, P.R., and Young, P.C. (1988) The effects of sulphur dioxide on phloem transport in two cereals, *Jnl. of Exp. Botany*, **39**, 997–1007.

Henderson-Sellars, B., Young, P.C., and Ribeiro da Costa, J. (1990) Water quality models: rivers and reservoirs, in *Proc. 1988 Int. Symp. on Water Quality Modelling of Agricultural Non-Point Sources*; United States Dept of Agriculture, Agricultural Research Service; 1990; Publication No. ARS-81, 381–420.

Jakeman, A.J., and Young, P.C. (1980) Towards optimal modelling of translocation data from tracer studies, *Proc. 4th Biennial Conf. Simulation Soc. of Australia.*, 248–253.

Jakeman, A.J., Littlewood, I.G., and Whitehead, P.G. (1990) Computation of the instantaneous unit hydrograph and identifiable component flows with application to two small upland catchments, *Jnl. of Hydrology*, **117**, 275–300.

Jury, W.A. (1982) Simulation of solute transport using a transfer function model, *Water Resources Research*, **18**, 363–368.

Jury, W.A., and Sposito, G. (1986) Field calibration and validation of solute transport models for the unsaturated zone, *Soil Sci. Soc. Amer. Jnl.*, **49**, 1331–41.

Kraijenhoff, D.A., and Moll, J.R. (1986) *River Flow Modelling and Forecasting*, (Water Science and Technology Library), D. Reidel, Dordrecht.

Ljung, L., and Soderstrom, T. (1983) *Theory and Practice of Recursive Estimation*, MIT Press, Cambridge, Mass.

Middleton, R.H., and Goodwin, G.C. (1990) *Digital Control and Estimation: A Unified Approach*, Prentice Hall, Englewood Cliffs, N.J.

Norton, J.P. (1986) *An Introduction to Identification*, Academic Press, New York.

Orlob, G.T. (ed.) (1983) *Mathematical Modelling of Water Quality: Streams, Lakes and Reservoirs*. J. Wiley, Chichester.

Rao, P.S.C., Ralston, D.E., Jessup, R.E., and Davidson, J.M. (1980) Solute transport in aggregated porous media: theoretical and experimental evaluation, *Soil Sci. Soc. Amer. Jnl.*, **40**, 1139–46.

Ribeiro da Costa, J. (1992) *Characterisation of Transport and Mixing Phenomena in the Ave River*, Ph.D. thesis, Centre for Research on Environmental Systems, University of Lancaster.

Skopp, J., and Warrick, A.W. (1974) A two phase model for miscible displacement of reactive solutes in soils, *Soil Sci. Soc. Amer. Proc.*, **38**, 545–550.

Smettem, K.R.J. (1984) Soil water residence time and solute uptake: 3. Mass transfer under simulated winter rainfall conditions in undisturbed soil cores, *Jnl. Hydrology*, **67**, 235–248.

Smith, R. (1987) A two equation model for contaminant dispersion in natural streams, *Jnl. of Fluid Dynamics*, **178**, 257–277.

Soderstrom, T., and Stoica, P. (1989) *System Identification* (Systems and Control Engineering Series), Prentice Hall, Hemel Hempstead.

Taylor, G.I. (1954) The dispersion of soluble matter in solvent flowing slowly through a tube, *Proc. Royal Soc. of London, Series A*, **223**, 446–468.

Wallis, S.G., Young, P.C., and Beven, K.J. (1989) Experimental investigation of the Aggregated dead zone model for longitudinal solute transport in stream channels, *Proc. Ins. Civ. Engrs (U.K.)*, **87**, (Part 2), 1–22.

Weyman, D.R. (1975) *Runoff Processes and Streamflow Modelling*, Oxford University Press, Oxford.

Whitehead, P.G., Young, P.C., and Hornberger, G.H. (1979) A Systems Model of Stream Flow and Water Quality in the Bedford–Ouse River, I: Streamflow Modelling, *Water Research*, **13**, 1155–1169.

Young, P.C. (1975) Recursive approaches to time-series analysis, *Bull. Inst. Math. and Applic.*, **10**, 209–224.

Young, P.C. (1981) Parameter estimation for continuous-time models: a survey, *Automatica*, **17**, 23–39.

Young, P.C. (1984) *Recursive Estimation and Time Series Analysis*, (Communication and Control Engineering Series), Springer-Verlag, Berlin.

Young, P.C. (1985) The instrumental variable method: a practical approach to identification and system parameter estimation, in: H.A. Barker and P.C. Young (eds) *Identification and System Parameter Estimation 1985, Vols 1 and 2*, Pergamon, Oxford, 1–16.

Young, P.C. (1986) Time-series methods and recursive estimation in hydrological systems analysis, in: D.A. Kraijenhoff and J.R. Moll (eds), *River Flow Modelling and Forecasting*, D. Reidel, Dordrecht.

Young, P.C. (1989) Recursive estimation, forecasting and adaptive control, in: C.T. Leondes (ed.), *Control and Dynamic Systems*, Academic Press, San Diego, 119–166.

Young, P.C. (1990) Systems methods in the evaluation of environmental pollution problems, in: R.M. Harrison (ed.), *Pollution – Causes, Effects and Control*, Royal Society of Chemistry: London, 367–388.

Young, P.C., and Beven, K.J. (1992) Comments on computation of the instantaneous unit hydrograph and identifiable component flows with application to two small upland catchments, forthcoming, *Jnl. Hydrology*.

Young, P.C., and Benner, S.V. (1991) *microCAPTAIN Handbook*, Version 2.0, Centre for Research on Environmental Systems, I.E.B.S., Univ. of Lancaster.

Young, P.C., and Jakeman, A.J. (1980) Refined instrumental variable methods of recursive time-series analysis, part 3: extensions, *Int. Jnl. Control*, **31**, 741–764.

Young, P.C., and Lees, M. (1992) The Active Mixing Volume: a new concept in modelling natural systems, in V. Barnett and F. Turkman (eds), *Statistics in the Environment*, J. Wiley, Chichester.

Young, P.C., and Minchin, P.E.H. (1991) Environmetric time-series analysis: modelling natural systems from experimental time-series data, *Int. Jnl. of Biological Macromolecules*, **13**, 190–201.

Young, P.C., and Wallis, S.G. (1985) Recursive Estimation: A Unified Approach to Identification, Estimation and Forecasting of Hydrological Systems, *Applied Mathematics and Computation*, **17**, 299–334.

Young, P.C., and Wallis, S.G. (1986) The Aggregated Dead Zone (ADZ) model for dispersion in rivers, *BHRA Int. Conf. on Water Quality Modelling in the Inland Natural Environment*, 421–433.

Young, P.C., and Wallis, S.G. (1992) Solute Transport and Dispersion in Channels, forthcoming in: K.J. Beven and M.J. Kirby (eds), *Channel Networks*, J. Wiley, Chichester.

Young, P.C., Chotai, A., and Tych, W. (1991) Identification, estimation and control of continuous-time systems described by delta operator models, in: N.K. Sinha and G.P. Rao (eds), *Identification of Continuous-Time Systems*, Kluwer Academic Publishers, Dordrecht, 363–418.

Young, P.C., Jakeman, A.J., and McMurtrie, R. (1980) An instrumental variable method for model order identification, *Automatica*. **16**, 281–294.

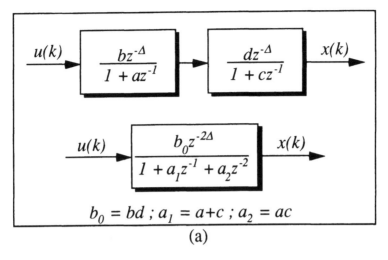

(a)

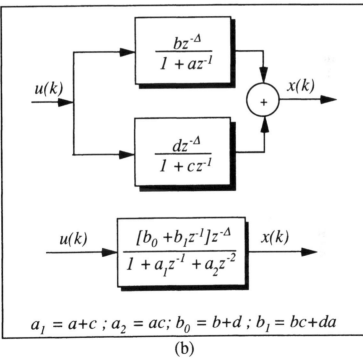

(b)

Figure 1   Serial and parallel connection of first order transfer function models. (a) serial connection; (b) parallel connection (single equivalent, second order block diagram shown below in each case)

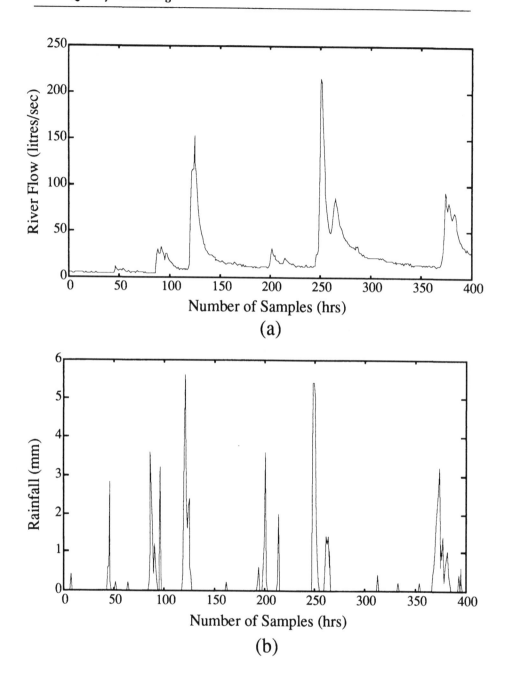

Figure 2 Rainfall-flow example I: rainfall-flow data from upland catchment. (a) hourly flow (discharge); (b) hourly rainfall over same time period as (a)

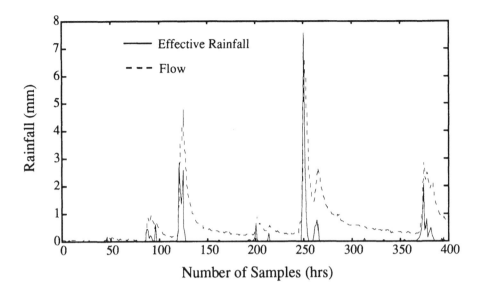

Figure 3 **Rainfall-flow example I: hourly effective rainfall obtained from the bilinear transform of the original rainfall series in Fig. 2(b), as proposed in the text, compared with flow from Fig. 2(a)**

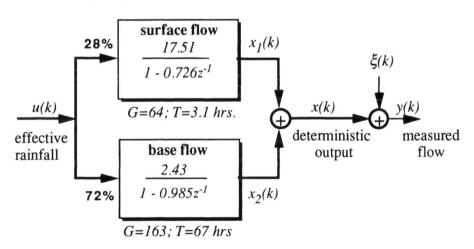

Figure 4 **Rainfall-flow example I: the z^{-1} operator *TF* model considered as a parallel connection of two first order processes (*G* denotes steady state gain; and *T* the time constant or residence time)**

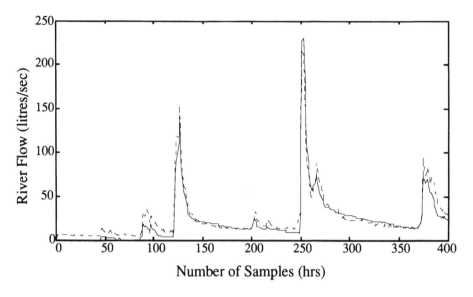

Figure 5 **Rainfall-flow example I: comparison of *SRIV* estimated delta operator model output (full) with measured output (dashed)**

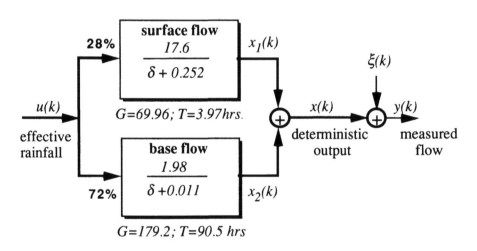

Figure 6 **Rainfall-flow example I: the delta operator *TF* model considered as a parallel connection of two first order processes (*G* denotes steady state gain; and *T* the time constant or residence time)**

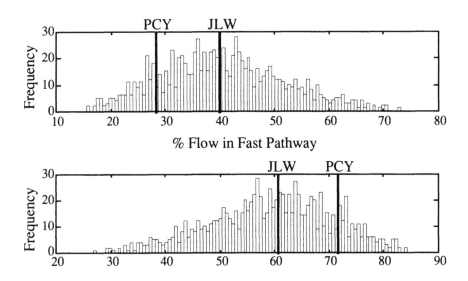

Figure 7 Rainfall-flow example I: Monte-Carlo evaluation of the JLW model (1000 realizations)

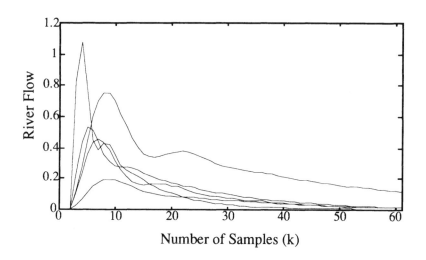

Figure 8 Rainfall-flow example II: superimposed plot of *micro-CAPTAIN* estimated hydrographs

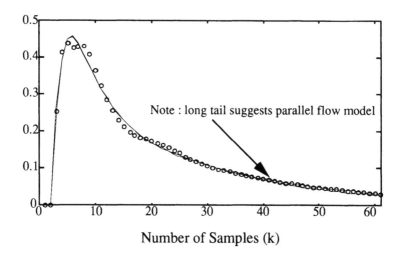

Number of Samples (k)

Figure 9   Rainfall-flow example II: Model of average hydrograph. (a) model (full); (b) average estimated hydrograph (o); (c) parallel approximation (fine – very similar to (a))

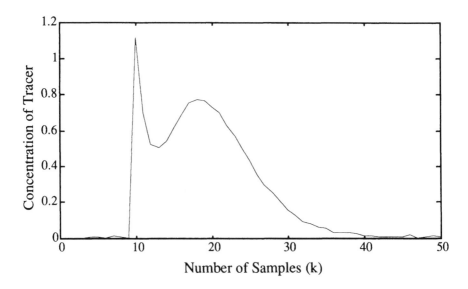

Number of Samples (k)

Figure 10   Soil tracer experiment: breakthrough curve for impulsive input of tracer

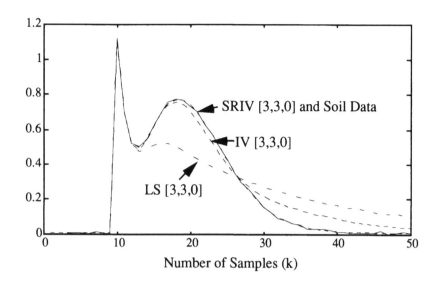

Figure 11 **Soil tracer experiment: comparison of different** *micro-CAPTAIN* **estimated models**

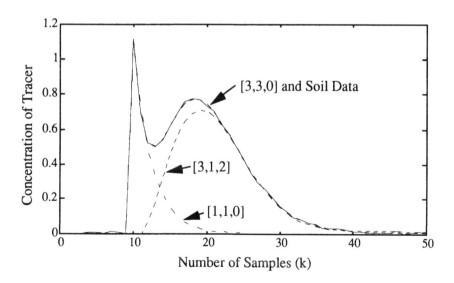

Figure 12 **Soil tracer experiment: model interpreted as parallel connection of a first order and a second order system**

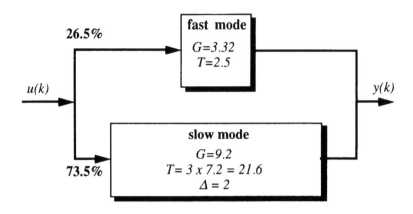

Figure 13 Soil tracer experiment: *TF* block diagram showing partitioning of water along parallel pathways (*G* denotes steady state gain; *T* the time constant or residence time; and Δ the pure time delay)

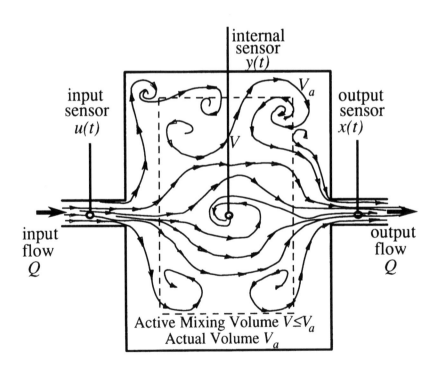

Figure 14 The Active Mixing Volume (AMV): an important new concept in environmental systems modelling?

3   The role of models in environmental impact assessment

Paul Whitehead

Introduction

There are a wide range of models that can be used for environmental impact assessment. These include:

(a)   time series techniques that result in input–output models which can provide information on the dynamic response of systems;
(b)   lumped models that characterize the dominant processes operating and are therefore to some extent physically based;
(c)   distributed models that may contain sophisticated descriptions of the processes and include detailed descriptions of the spatially-distributed nature of the system.

The distributed models often appear attractive for EIA analysis but are generally hampered by the needs for distributed information and large numbers of parameters. Such information is frequently unavailable for environmental impact studies. However, theoretical studies using such models is often of value in EIA.

In this paper three examples of models used for EIA are given. These include MAGIC (Model of Acidification of Groundwaters in Catchments), QUASAR (Quality Simulation Along Rivers) and a nitrogen simulation model for reservoirs.

The MAGIC model

MAGIC (Model of Acidification of Groundwater in Catchments; Cosby et al, 1985,1986) is explicitly designed to perform long term simulations of changes in soil water and streamwater chemistry in response to changes in acidic deposition. The processes on which the model is based are:

(a) hydrological and chemical mass balance;
(b) anion retention by catchment soils (e.g. sulphate adsorption);
(c) adsorption and exchange of base cations and aluminium by soils;
(d) alkalinity generation by dissociation of carbonic acid (at high CO_2 partial pressures in the soil) with subsequent exchange of hydrogen ions for base cations;
(e) weathering of minerals in the soil to provide a source of base cations;
(f) control of Al^{3+} concentrations by an assumed equilibrium with a solid phase of Al (OH).

A sequence of atmospheric deposition is assumed for MAGIC and current deposition levels of base cations, sulphate, nitrate and chloride are needed along with some estimate of how these levels had varied historically. Historical deposition variations may be scaled to emissions records or may be taken from other modelling studies of atmospheric transport into a region. Weathering estimates for base cations are extremely difficult to obtain. Nonetheless, it is the weathering process that controls the long-term response and recovery of catchments to acidic deposition and some estimate is required. An optimization procedure is used to calibrate the model to individual catchments and details of the model equations and application are given elsewhere (Cosby et al, 1985, 1986; Wright et al, 1986).

Comparisons with palaecological data

In order to obtain confidence in a model such as MAGIC it is necessary to compare the results with other data indicating long-term acidification.

Unfortunately river quality records have been shown to be inadequate for this analysis (Warren et al, 1986) and there are few records of soil chemistry that contain accurate historical records. However, the palaecological records from lake sediments (Battarbee et al, 1985; Flower et al, 1983) do provide an assessment of pH changes over a long period of time. For example, Figure 1 shows the pH trends for Round Loch of Glenhead and indicates a decline of 1 pH unit from 1850 to

1980. The figure also shows the simulated pH levels obtained using MAGIC applied to the Round Loch of Glenhead. A parameter sensitivity analysis indicates the confidence bounds associated with the simulated trend.

Assessing future trends and land use change effects

In order to investigate the effects of changes in atmospheric pollution levels and afforestation MAGIC has been applied to a sensitive catchment in the Galloway Region of South West Scotland. The particular catchment is Dargall Lane in the Loch Dee system and extensive monitoring and data collection and analysis has been undertaken by the Solway River Purification Board. A detailed description of the Loch Dee study is given by Burns et al (1984) and details of the MAGIC application to Dargall Lane are presented by Cosby et al (1986a).

Figure 2 shows a simulation of long-term acidity for the Dargall Lane catchment. The historical simulation of pH shown in Figure 2 is similar to the values obtained from the diatom records of lochs in the region in that a significant decrease in pH from 1900 onwards is inferred. The steeper decline from 1950 to 1970 follows from the increased emission levels during this period. The model can also be used to predict future stream water acidity given different future deposition levels. For Dargall Lane stream acidity trends are investigated assuming two scenarios for future deposition. Firstly assuming deposition rates are maintained in the future at 1984 levels, the model indicates that annual average stream pH is likely to continue to decline below presently measured values. Second, assuming deposition rates are reduced by 50% from 1984 levels (between 1985 and 2000) the results indicate that current stream water acidity will be maintained (Figure 2). Further details of the application of this model are given elsewhere (Cosby et al, 1986). Note an increase in stream water pH in about 1980; this follows a significant drop in sulphur emissions during the 1970s. Note also that an earlier decline in streamwater acidity is predicted if there had been no reductions in emissions since 1970.

Afforestation

Afforested systems are more complex to model than grassland systems because the introduction of the forest perturbs a grassland ecosystem which in itself is difficult to model. The effects of the forest root system, leaf litter layer and drainage ditches will change the hydrological pathways; this will control the nature and extent of the chemical

55

reactions in the soil and bedrock. Further, the additional filtering effect of the trees on the atmosphere will enhance occult/particle deposition and evapotranspiration will increase the concentration of dissolved components entering the stream. The magnitude of these different effects varies considerably; for example evapotranspiration from forests in the British Uplands is typically of the order of 30% of the precipitation which is almost twice the figure for grassland. This will have the consequence that the total anion concentrations within the stream and soil waters increase by 14% following afforestation. The forest will also increase anion and cation loading due to the enhanced filtering effect of the trees on air and occult sources. The filtering effects will apply both to marine and pollutant aerosol components. Altering the hydrological pathways can also have a major effect on stream water quality since the forest tends to increase surface runoff thereby flushing/displacing highly acidic water from the surface layers; the soil zone acts as a proton and aluminium source whilst the bedrock, if silicate or carbonate bearing, provides proton consumption by weathering reactions. To illustrate the effects of afforestation simply in terms of increased concentrations from both enhanced dry deposition and evapotranspiration, the MAGIC model has been applied to the Dargall Lane catchment assuming that a forest is developed over the next 40 years. It should also be noted that, here, no allowance has been made for the effects of cation and anion uptake by the trees during their development; the incorporation of base cations into the biomass would result in an enhanced acidification effect during this period.

Of critical importance is the relative and absolute contribution of marine and pollutant inputs from dry and occult deposition. Figure 3 A-C shows the effects of increasing evapotranspiration from 16% to 30% over the forest growth period with varying levels of marine, pollutant and marine + pollutant inputs. Increasing either marine or pollutant components leads to enhanced stream water acidity, the greatest effects being observed when both components are present; the effect of simply increasing evapotranspiration from 16% to 30% has a similar effect but the changes are much smaller. The important features of these results are that the enhanced and acidic oxide inputs from increased scavenging by the trees results in a marked reduction in pH levels and that there is an additive effect when both processes are combined. These reductions are much greater than the effect of evapotranspiration.

Pollution 'climate' effects

An important factor in determining stream acidity in the upland UK is the level of acidic oxide deposition; rates of deposition (non marine wet deposition and dry deposition) can vary from $0.5g$ N $m^{-2}y^{-1}$. Figure 4 A-B shows the effects of such variations for both moorland and forested catchments; the highest levels correspond to areas with high atmospheric acidic oxide rates (3 times the historic and 1984 deposition levels observed in Southern Uplands of Scotland). With increasing atmospheric acidic oxide pollution, the decline in stream pH is accelerated, the changes occur much earlier, and the final pH of the stream water is lower.

Thus MAGIC can be used to investigate the impact of atmospheric pollution and afforestation in sensitive upland catchments.

River quality modelling – The QUASAR model

The water quality model QUASAR has been designed at the Institute of Hydrology to assess the impact of pollutants on river systems. The model was originally developed as part of the Bedford Ouse Study, a DOE and Anglian Water Authority funded project initiated in 1972. The primary objective was to simulate the dynamic behaviour of flow and water quality along the river system (Whitehead et al, 1979, 1981). Initial applications involved the use of the model within a real time forecasting scheme collating telemetered data and providing forecasts at key abstraction sites along the river (Whitehead et al, 1984).

The model was also used within a stochastic or Monte-Carlo framework to provide information on the distribution of water quality within river systems, particularly in rivers subjected to major effluent discharges (Whitehead and Young, 1979). This technique was also used by Warn and Brew (1980) and Warn and Matthews (1984) to assess mass balance problems within river systems. There has also been a range of model applications to other UK rivers such as the Tawe in South Wales to assess heavy metal pollution, and the Thames to assess the movement and distribution of nitrates and algae along this river system (Whitehead and Williams, 1982; Whitehead and Hornberger, 1984).

A total of eight water quality variables are simulated in addition to flow, including nitrate, ionized and unionized ammonia, dissolved oxygen (DO), biochemical oxygen demand (BOD), nitrate, ammonia, pH, temperature, *E.coli* and any conservative pollutant or inert material in solution. A wide range of inputs can be investigated including tributaries, groundwater inflows, direct runoff, effluents, storm water and the model can allow for abstractions for public water supply or irrigation. A multi-reach approach is utilized so that the user specifies reach boundaries or locations of primary interest.

57

The model can be used deterministically or stochastically and is essentially physically based with some empirical formulations such as velocity-flow relationships. In the deterministic or dynamic mode pulses of pollutant can be traced downstream, while in the stochastic or planning mode, effluent consent conditions can be established given a river quality objective.

Model description

A detailed description of the underlying model structure is given by Whitehead et al (1979, 1981). Essentially the model performs a mass balance on flow and water quality sequentially down a river system. The model takes into account inputs from tributaries, groundwaters and effluent discharges, and allows for abstractions, chemical decay processes and biological behaviour along the river system. In addition the model accounts for the varying travel times operating at differing flow conditions so that pollutants are transported along the river at realistic velocities. The model operates in two modes; planning mode (stochastic simulation) and operational mode (dynamic simulation).

Planning (or stochastic) mode

In the planning or stochastic mode a cumulative frequency curve and distribution histogram of a water quality parameter are generated by repeatedly running the model using different input data selected according to probability distributions defined for each input variable. Whitehead and Young (1979) and Warn and Brew (1980) have used this technique, known as Monte-Carlo simulation, to provide information for long-term planning and water quality management. In this mode statistical data of the water quality and flow in the first reach at the top of the river, and in tributaries, discharges and abstractions are required. These data include, for each variable input to the model, the mean, standard deviation and distribution type, i.e. lognormal, rectangular or gaussian. Random numbers are generated and water quality and flow values are then chosen from these characterized distributions. A mass balance is performed at the top of each reach to include tributaries, discharges and any other inputs to the river. The values generated by the model equations represent the water quality or flow at the end of the reach. The model equations are solved using random numbers as the input values until steady state has been reached. Five hundred and twelve random simulations are generated. The output is stored and used to produce cumulative frequency distributions and distribution histograms (see Figure 5).

Operational (or dynamic) mode

In the operational or dynamic mode, water quality and flow are simulated over selected periods. This enables the possible effects of a pollution event on a river to be investigated. In this mode time series data are required for water quality and flow parameters for the first reach of the river and for tributaries, effluent discharges and abstractions along the section of the river of interest. A mass balance is performed over each reach of the river and a time series of downstream flow and quality computed. A typical model output is given in Figure 6. The model has been set up in a real time operational model for the whole of the Great Ouse catchment and used to provide forecasts to operational managers during pollution events (Whitehead et al, 1984).

Data requirements

Three sets of data are required by QUASAR; a catchment structure consisting of a river map, boundary conditions which define the water quality and flow of the tributaries and of the water at the top of the river, and reach parameters consisting of data specific to each reach. QUASAR has been applied to several environment impact studies including the Roadford Reservoir Scheme and the Thames Maidenhead Flood Relief Channel and is available from the Institute of Hydrology.

Modelling nitrate in reservoirs

Sudden increases in nitrate in river systems can significantly affect reservoir systems leading to the possibility that public water supplies could be in excess of the EC directives for nitrate in drinking water. A model of a reservoir can assist in assessing the impact of a nitrate flush and improved operational management practices. In most reservoir systems, ammonia concentrations are low and the dominant nitrogen transformation process is denitrification. An input output mass balance model should allow for given inflow and outflow pumping rates and concentrations and should account for the denitrification process. The basic budget for a well-mixed reservoir can be described in the following equation:

$$V \frac{dC}{dt} = C_r Q_{in} - C Q_{out} - \frac{K_4}{d} . 1.047^{(T-20)} C \tag{1}$$

where d is the depth of the reservoir, C and C_r are the concentrations of nitrate in the mixed reservoir and river respectively, Q_{in} and Q_{out} are the input and output pumping rates, and K_4 is the denitrification rate (which is a function of temperature $T°C$).

Equation (1) can be solved on a weekly basis by numerical integration using either observed weekly nitrate data or hypothetical values to test different operating strategies.

The model has been set up for the Farmoor, Queen Mary and Queen Mother Reservoirs using data from 1980 and 1979 respectively and details of these applications are given by Whitehead and Toms (1991). Figure 7 shows the model applied to Farmoor I over some 12 years of weekly data. In general an excellent fit to the data is obtained. Also presented in Figure 7 is the volume of water in the reservoir expressed as % full and the mean residence time. The reservoir is kept full for much of the time especially in the early years of operation although droughts in years 1971, 1973 and 1975/76 show reservoir volumes dropping. The mean residence time is 24 weeks in the first six years of operation but in the later years this falls to approximately ten weeks. This period corresponds with increasing nitrate levels in the reservoir which appears due to a combination of higher river nitrate concentrations and decreased reservoir residence times.

In using equation (1) to fit calculated to observed data, real values of flow rates and river concentrations were used. It is also possible to use the model to determine reservoir response to a variety of hypothetical or predicted river nitrate patterns. An example of such a predictive exercise is shown in Figure 8, where it has been assumed that the reservoir has a retention time of 100 days and the regular annual nitrate cycle has a constant maximum of 12 mg l^{-1} during the winter quarter and a constant minimum of 6 mg l^{-1} during the summer quarter. The reservoir is assumed to be well-mixed, and its predicted nitrate concentration with no denitrification (i.e. $K_4 = 0.0$) is shown to match the average in the river as would be expected. However, the maximum nitrate concentration is lower than that in the river, and delayed for 2 months: this illustrates the protection afforded by bank-side storage derived purely from hydraulic effects. The other graphs in Figure 8 show what might happen if the reservoirs exhibited denitrification with a typical K_4 of 0.06 m.day^{-1}, and had mean depths of 20m and 10m respectively. In the deeper reservoir denitrification removes a 0.9 mg l^{-1} from the maximum whilst, in the shallower reservoir, 1.5 mg l^{-1} of nitrate nitrogen is removed. This indicates that, for a given retention time, shallower reservoirs or lakes are more efficient at removing nitrate-nitrogen. From such studies it is possible to identify optimal conditions for denitrification. These conditions may be summarized as follows:

(a) shallow water (maximum surface area of mud to water volume);
(b) no stratification (warmer water against the sediment);
(c) long retention time (maximum time for denitrification to occur);
(d) high primary production (concomitant low populations of zoo-plankton);
(e) no exposure of sediments during draw down.

However, many of these optimal conditions cannot be sustained because of conflicting water resource requirements – namely the need to maintain reservoirs at the maximum level for adequate water supply. However, the most important condition, that of maximizing residence time, is consistent with maintaining reservoirs at near full and will meet both objectives. Unfortunately, demands on water are such that major reservoir drawdowns are anticipated in dry summers, and the best strategy is to avoid using poor quality river water high in nitrates. This could be achieved by avoiding the first flush of nitrates following a relatively dry spell.

Conclusion

In order to assess environmental impact it is not necessary to use particularly complex models. Indeed such models may lead to incorrect conclusions because they are being used outside of their range of acceptable application or are based on guessed parameter values. The lumped model approach offers a pragmatic approach in which initiative, skill and experience can be used to assess model behaviour and the uncertainties associated with incomplete knowledge can be assessed.

References

Battarbee, R.W., Flower, R.J., Stevenson, A.C. and Ripper, B. (1985) A palaeoecological test of competing hypotheses, *Nature*, **314** (6009): 350–352.

Burns, J.C., Coy, J.S., Tervet, D.J., Harriman, R., Morrison, B.R.S. and Quine, C.P. (1984) The Loch Dee project: a study of the ecological effects of acid precipitation and forest management on an upland catchment in South West Scotland, *Fish. MANAGE*, **15**: 145–167.

Cosby, B.J., Wright, R.F., Hornberger, G.M. and Galloway, J.N., (1985a) Modelling the effects of acid deposition: Estimation of long term water quality responses in a small forested catchment, *Water Resources Res.*, **21**, No. 1, 1591–1606.

Cosby, B.J., Hornberger, G.M., Galloway, J.N. and Wright, R.F. (1985b) Modelling the effects of acid deposition: Assessment of a lumped parameter model of soil water and stream water chemistry, *Water Resources Res.*, **21**, No. 1, 51–53.

Cosby, B.J., Whitehead, P.G. and Neal, C.R. (1986) A preliminary model of long term changes in stream acidity in South West Scotland, *J. Hydrol.*, **84**: 381–401.

Flower, R.J. and Battarbee, R.W. (1983) Diatom evidence for recent acidification of two Scottish lochs, *Nature*, **3051** (5930): 130–133.

Warn, A.E. and Brew, J.S. (1980) Mass Balance, *Water Res.*, **14**: 1427–34.

Warn, W.E. and Matthews, C. (1984) Calculation of the compliance of discharges with emission standards, *Wat. Sci. Tech.* **16**: 183–96.

Warren, S.C. et al (1986) Acidity in UK Fresh Waters, *UK Acid Waters Review Committee Report*, DOE, April.

Whitehead, P.G. and Young, P.C. (1979) Water quality in river systems, Monte-Carlo analysis, *Water Resources Res.*, 451–459.

Whitehead, P.G. and Hornberger, G.M. (1979) A systems model of stream flow and water quality in the Bedford Ouse river – I. Stream flow modelling, *Water Res.*, **13**: 1155–69.

Whitehead, P.G., Beck, M.B. and O'Connell, E. (1981) A systems model of flow and water quality in the Bedford Ouse river system – II. Water quality modelling, *Water Res.*, **15**: 1157–71.

Whitehead, P.G. and Williams, R.J. (1982) A dynamic nitrate balance model for river basins, *IAHS Exeter Conference Proceedings*, IAHS publication no. 139.

Whitehead, P.G., Caddy, D. and Templeman, R. (1984) An on-line monitoring, data management and forecasting system for the Bedford Ouse river basin, *Wat. Sci. Tech.* **16**: 295–314.

Whitehead, P.G. and Hornberger, G.M. (1984) Modelling algal behaviour in the river Thames, *Water Res.*, **18**, no. 8: 945–53.

Whitehead, P.G. (1990) Modelling nitrate from agriculture into public water supplies, *Phil. Trans. R. Soc. Lond.*, **B 3229**, 403–410.

Whitehead, P.G. and Toms, I.P. (1991) A dynamic model of nitrate in reservoirs (in press).

Wright, R.F., Cosby, B.J., Hornberger, G.M. and Galloway, J.N. (1986) Comparison of palaeolimnological with MAGIC model reconstruction of water acidification, *Water Air and Soil Pollution*, **30**: 307.

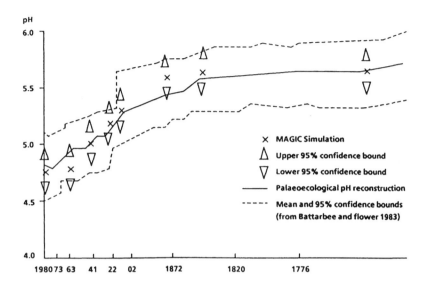

Figure 1 pH reconstruction and MAGIC simulation for Round Loch of Glenhead

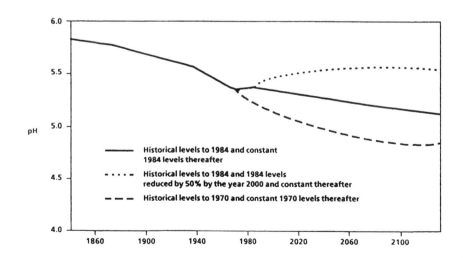

Figure 2 Simulated historical and future pH at Dargall Lane

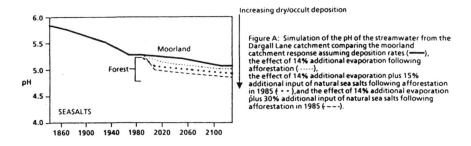

Figure A: Simulation of the pH of the streamwater from the Dargall Lane catchment comparing the moorland catchment response assuming deposition rates (——), the effect of 14% additional evaporation following afforestation (·······), the effect of 14% additional evaporation plus 15% additional input of natural sea salts following afforestation in 1985 (· ·), and the effect of 14% additional evaporation plus 30% additional input of natural sea salts following afforestation in 1985 (– – –).

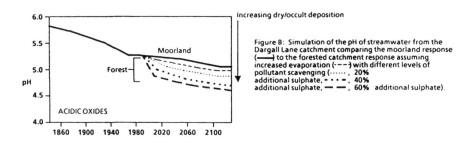

Figure B: Simulation of the pH of streamwater from the Dargall Lane catchment comparing the moorland response (——) to the forested catchment response assuming increased evaporation (– – –) with different levels of pollutant scavenging (·······, 20% additional sulphate, · · · , 40% additional sulphate, — — , 60% additional sulphate).

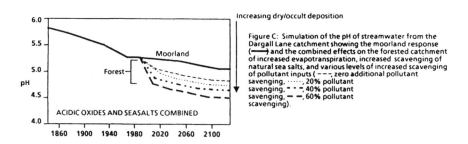

Figure C: Simulation of the pH of streamwater from the Dargall Lane catchment showing the moorland response (——) and the combined effects on the forested catchment of increased evapotranspiration, increased scavenging of natural sea salts, and various levels of increased scavenging of pollutant inputs (– – –, zero additional pollutant savenging, ·······, 20% pollutant savenging, · · · , 40% pollutant savenging, — —, 60% pollutant scavenging).

Figure 3   Combined effects of afforestation and acid deposition on pH at Dargall Lane

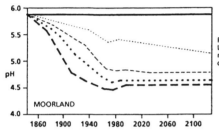

Figure A: Simulation of the pH of streamwater from the Dargall Lane moorland catchment assuming sulphate deposition patterns modified by various factors to reproduce a range of loading conditions (ie from pristine to heavy pollution).

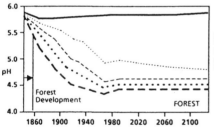

Figure B: Simulation of the pH of streamwater from the 'forested' Dargall Lane catchment assuming afforestation from 1884 onwards and sulphate deposition patterns multiplied by various factors to reproduce a range of loading conditions from pristine to heavy pollution.

Figure 4   Impact of afforestation under varying pollution climates

65

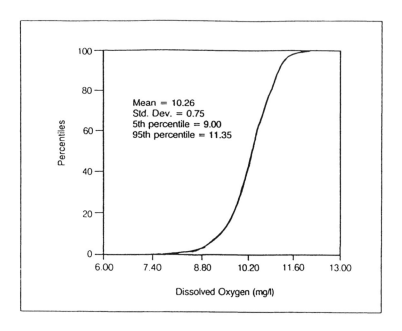

Figure 5 Output from QUASAR in planning mode showing distribution of water quality

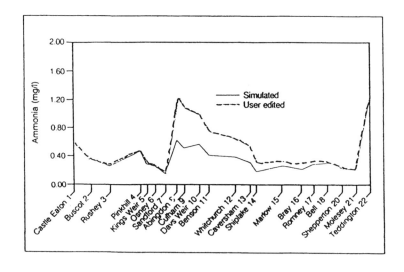

Figure 6 Output from QUASAR in operational mode

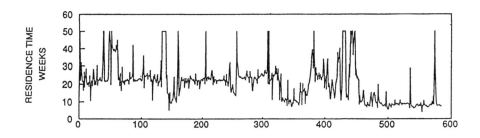

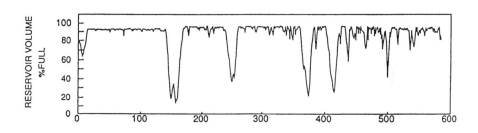

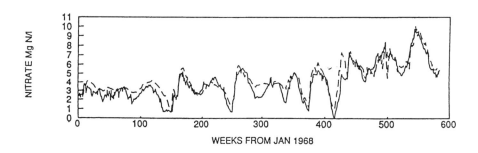

Figure 7 Farmoor Reservoir nitrate, level and residence time simulations

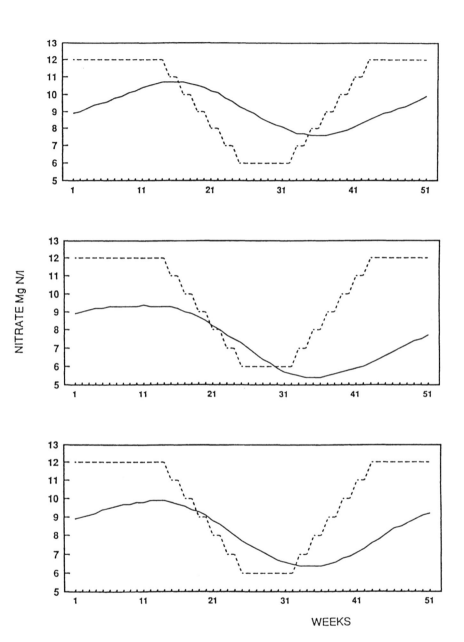

Figure 8 Simulated reservoir response for different K and depth values. Top graph = 0.0 m.day^{-1} depth = 10m; Middle graph K$_4$ = 0.06 m.day^{-1}, depth = 10m; Lower graph K$_4$ = 0.06 m.day^{-1}, depth = 20m

4　Modelling river water quality and impact from sewers and storm sewer overflows

Anders Malmgren-Hansen and Hanne K. Bach

Abstract

Sewer and storm sewer overflows have a well recognized detrimental effect on river water quality downstream of urban areas. The need for consent setting based on environmental legislation emphasizes the need for models which relate the water quality and ecological state of a river to actual and future planned loadings.

Many of the physical, chemical and biological processes determining the effect of sewer and storm sewer overflows on river water quality are well-known and described in model formulations. The integration of these process descriptions with dynamic hydraulic models has enhanced the usefulness of river water quality models.

The availability of powerful personal computers and workstations along with advanced software utilities will support the further development of user-friendly modelling tools covering the integrated effects of water quality parameters such as oxygen, ammonia, pH, metal-compounds and suspended sediments on the aquatic life and ecological state of rivers.

Key words: River water quality modelling. Sewer and storm sewer overflows. Model formulations. Modelling tools.

Introduction

A wide range of water quality problems and the deterioration of the ecological state in rivers are caused by inflows from sewers and storm sewer overflows. This had already created major problems at the beginning of this century especially with respect to oxygen conditions. The pioneering work by Streeter and Phelps who developed the first water quality model in 1925 to describe the oxygen depletion in the Ohio river, is well known.[1] An early and well known action concerning river water quality impact from sewers has been the clean-up of the river Thames.

Besides oxygen depletion, bacterial pollution was recognized early on as a major public health problem especially in relation to the use of river water in water supply and the consumption of fish and shellfish caught in rivers and estuaries. These are still major problems.[2] Other problems are related to toxicity of non-ionized ammonia, toxic chemicals and heavy metals discharged from urban areas.

Suspended matter discharged from sewers and storm sewers is known to change the sediment conditions in rivers affecting the living conditions for benthic plants and animals. In smaller rivers the periodic increases in discharges from storm sewers wash out organisms downstream of the discharge.[3]

The increased public awareness of the quality of water and ecology in rivers has resulted in efforts in many countries to regulate the urban discharges and to classify the extent to which the rivers are affected.[4,5,6]

The EC directives coming concerning bacterial pollution and waste water,[7,8] and the directive concerning the ecological state of the aquatic environment, which is under preparation, reflects the increasing focus on environmental problems.

The need for consent settings based on the present and expected directives and legislation emphasizes the need for methods suitable for establishing the relation between the varying effluent quantities and pollutants from sewers and storm sewers and the resulting river water quality and ecology. The answer to this demand is the use of models which relate the statistics of effluents to the environmental state in the river and which also take into account the hydrological and hydraulic conditions.

In the following the physical, chemical and biological processes which influence the water quality and ecological conditions in rivers are described according to their usual representation in mathematical models.

These processes are later related to the present modelling capabilities using existing modelling tools.

70

Processes affecting river water quality and ecological state

A major negative effect of sewers and storm sewer overflows on the water quality in rivers is the reduction in oxygen content. At the same time the oxygen conditions are the most important factors affecting the aquatic life in rivers. The most important water quality standards are therefore related to the concentration of dissolved oxygen and parameters affecting this.

Degradation of organic material

The oxygen consumption from degradation of organic material is normally measured as Biochemical Oxygen Demand $\approx$ BOD. BOD_5 is the amount of oxygen consumed within 5 days. Normally the oxygen consumption is described as a 1st order decay:

$$\frac{dL}{dt} = -K_1 \cdot L \tag{1}$$

where

L	is the concentration of BOD (M/L^3)
K_1	is the 1st order decay constant $(1/T)$
t	is time (T)

As the biodegradable organic material is measured in oxygen equivalents its influence on the oxygen concentration is consequently described as:

$$\frac{dC}{dt} = -K_1 \cdot L \tag{2}$$

where

C is the oxygen concentration (M/L^3)

The biochemical oxygen demand may be separated into oxygen consumption from degradation of organic material and nitrification of ammonia. The oxygen consumption from nitrification of ammonia may then be described as:

$$\frac{dC}{dt} = 4.57 \cdot K_{1,NH4} \cdot S_{NH4} \tag{3}$$

71

where

4.57   is equal to the number of grams of oxygen necessary for the transformation of 1 g NH_4–N into nitrate

$K_{1,NH4}$ is the 1st order nitrification rate (1/T)

S_{NH4} is the concentration of ammonia (M/L^3)

The oxygen consumption rate is normally considered to be dependent on the water temperature following an Arrhenius-type law, e.g.:

$$K_1 = (K_1)_{20} \cdot \theta^{(T-20)} \tag{4}$$

where

T is the temperature of the water (°C)

$(K_1)_{20}$ is the value of K1 at 20°C (1/T)

θ is a dimensionless parameter

Re-aeration

Normally re-aeration is described as occurring at the air-water interface following a 1st order kinetic law:

$$\frac{dR_0}{dt} = K_2 (C_s - C) \tag{5}$$

where

dR_0/dt is the re-aeration rate ($M/L^3 \cdot T$)

C_s is the oxygen saturation constant (M/L^3)

C is the oxygen concentration in the river (M/L^3)

K_2 is the re-aeration kinetic constant (1/T)

The re-aeration kinetic constant is influenced by many factors. The most important considered are the water temperature, river flow characteristics (flow velocity, water depth, slope of river) and meteorological conditions (air temperature and wind). A large number of relationships between these factors and K_2 has been established.[9]

Photosynthesis and respiration

The photosynthetic oxygen source in streams and rivers from bottom vegetation is normally adjusted to a half-cycle sine wave following the sinusoidal variation in light intensity.[10]

$$P = P_m \sin (\pi/s) (t-t_s) \text{ when } t_s \le t \le t_s + s$$

$$P = 0 \qquad\qquad \text{when } t_s + s \le t \le t_s + 1$$

(6)

where

P_m	is the maximum oxygen production by photosynthesis $(M/L^3 \cdot T)$
t_s	is the time, expressed as the fraction of the day, at which photosynthesis becomes active
s	is the fraction of the day during which photosynthesis is active

When bottom vegetation is present in a stream or river it not only produces oxygen during the day, but it has a simultaneous oxygen uptake during both night and day. This respiration rate R, $(M/L^3 \cdot T)$ is normally considered to be constant during simulations. Special techniques may be adapted to estimate the size of P_m, R and the K_2 value.[11,12]

Sediment oxygen demand

Removal of organic material by sedimentation and/or adsorption were first described by Dobbins in 1964 as a 1st order process.[13] Furthermore he included the oxygen uptake by benthic processes such as the bacterial degradation of organic material naturally produced in the river system.

Taking into account sedimentation and/or adsorption, equation (1) becomes:

$$\frac{dL}{dt} = - (K_{1,s}) \cdot L$$

(7)

where

$K_{1,s}$ is the sedimentation/adsorption constant $(1/T)$

Taking into account re-aeration and the oxygen consumption by the sediments, equation (2) becomes:

$$\frac{dC}{dt} = - K_1 \cdot L + K_2(C_s - C) - D_b$$

(8)

where

D_b is the rate of oxygen removal caused by benthic processes

73

The first author to take into account the actual amount of BOD sedimented/adsorbed at the bottom was Harremoës in 1982, when he distinguished between immediate and delayed oxygen depletion.[14] The significance of this concept, especially when taking into consideration the immediate and delayed impact on the oxygen conditions in rivers from storm sewer overflows with a high content of readily settleable suspended organic content, was demonstrated by Hvitved-Jacobsen.[15]

When taking into account delayed oxygen demand, the BOD content in the water is separated into two components:

$$L = L_d + L_s \tag{9}$$

where

L_d is the dissolved fraction (M/L^3)

L_s is the suspended fraction which contributes to delayed oxygen demand (M/L^3)

When the suspended BOD fraction L_s deposits, an equivalent quantity of bottom sediment builds up:

$$\frac{dL_b}{dt} = \frac{dL_s}{dt}\frac{1}{h} = K_5 \cdot L_s \cdot \frac{1}{h} \tag{10}$$

where

h is the water depth (L)

L_b is the amount of sedimented BOD on the bottom (M/L^2)

K_5 is the sedimentation rate of suspended BOD (M/T)

The sedimented BOD normally has a different degradation rate from BOD in the water phase and is normally described as a 1st order reaction:

$$\frac{dL_b}{dt} = -K_4 \cdot L_b \tag{11}$$

where

K_4 is the 1st order decay constant (1/T)

If steady state conditions are assumed, it is still possible to describe the oxygen conditions in rivers using analytical solutions to equations (1) to (11).[14]

In nature it is not always adequate to assume steady state conditions when describing the oxygen conditions in rivers. The first recognition of the possible importance of immediate oxygen consumption when sediments are resuspended in rivers was reported in 1979 by Meierholtz et al.[16] Resuspension is, by nature, an unsteady event, occurring at high flow velocities caused by e.g. storm sewer overflows, floods or the tidal influence in lower parts of the rivers and in estuaries.

As a first approximation resuspension of sedimented organic material is normally described as a zero order reaction occurring at flow velocities above a 'critical' value. If resuspension is also taken into account and integrating equations (10) and (11) one obtains:

$$\frac{dL_b}{dt} = K_5 \cdot L_s \cdot \frac{1}{h}$$

$$-K_4 \cdot L_b \tag{12}$$

$$- S_1 \cdot L_b \cdot \frac{1}{h}$$

where S_1 is the resuspension rate (M/T).

This simple description of sedimentation and resuspension is often inadequate and should be refined. Much research has already been carried out in describing the transport of organic material with cohesive characteristics. Many processes actually influence the erosion, transport and deposition of fine, cohesive material. Key processes are vertical mixing, flocculation, segregation, fluid development and consolidation.

The properties of cohesive materials have been studied in the laboratory, e.g. the consolidation of soft mud,[17] and implemented in integrated cohesive sediment transport models.[18,19]

pH variations

Oxygen concentrations are the most important parameter for the aquatic life and the ecological state in rivers,[20,21] but other factors are also of importance. In non-ionized form ammonia (NH_3) is toxic to fish, and the toxicity is not only dependent on the concentration of non-ionized ammonia but also on the simultaneous oxygen concentration.[22,23] The dissociation of ammonia into ionized and non-ionized form is highly dependent on the actual pH value in the river, with an increasing fraction being non-ionized at higher pH value. At pH values above 9 almost all ammonia is in a non-ionized form. In smaller rivers with a high photosynthetic activity, pH is known to increase due to the consumption of carbon dioxide resulting in potential toxic effects of non-ionized ammonia.[24,25]

Another problem related to pH is increasing acidification. Low pH values alone have an impact on the ecology in rivers. Furthermore, low pH values increase the solubility of alluminium which is toxic to the aquatic life in rivers.[26,27]

Other factors affecting river ecology

Dredging of sediments, e.g. in reservoirs, has caused drastic fish mortalities in downstream rivers. These mortalities have been correlated with the increase in the suspended solids and ammonia concentrations in the water. The high concentrations of suspended material have a combined effect with ammonia and lowered oxygen concentrations.[28]

Direct effects of eutrophication have, in certain cases, resulted in the growth of phytoplankton in river systems like the Loire river causing water quality and ecological problems beyond what is normally seen in rivers.[29]

A direct physical impact on the river ecology due to the increased flow from sewer storm overflows which cause the flushing out of invertebrates living downstream has also been reported.[3]

Many other factors may affect the river water quality and ecology including drainage, construction of reservoirs, extraction of water for water supply and irrigation. Such effects must also be taken into consideration in connection with sewer or storm sewer discharges.

Present state of modelling capabilities

The capabilities of modelling river water quality affected by sewers and storm sewer overflows have today reached a rather sophisticated level. Thanks to the availability of personal computers 4th generation tools are now available for non-expert users. In the following, a number of available modelling tools are mentioned including both tools suitable for generation of input data to the river models and a number of river models themselves.

Urban drainage models

Inputs from urban sewer systems to the river water quality models are often generated using a combination of actual measurements and results from urban drainage models. These models are designed to describe the flow and the contents of pollutants in the sewers and storm sewer overflows. Normally the pollutants described include biochemical oxygen demand and/or chemical oxygen demand such as in the FLUPOL[30] and the MOUSE[31] systems. Suspended solids are also included in the MOSQITO system.[32]

River water quality models

A large number of river water quality models have been developed since Streeter and Phelps gave their first and very elegant contribution in 1925.[10,13,14,24] All the models mentioned here assume steady state hydraulic conditions to enable an analytical solution of the mathematical model to be obtained.

The availability of computers has enabled the development of numerical models. Examples of numerical modelling systems developed in the seventies describing river water quality are QUAL II[33] and SYSTEM 11.[34,35] Besides these 3rd generation[36] general modelling systems designed for main frame computers, a number of more specialized models have been developed to be used on main frames or personal computers. A number of river quality models such as TOMCAT, SIMCAT and STREAM have been developed by water authorities through the eighties for consent setting in England and Wales.[37]

River water quality modelling systems are now developed as 4th generation tools to be used by external users on personal computers. Examples of models describing intermittent discharges are DOSMO[38] and SPRAT.[39] These models describe the immediate effects, and in DOSMO the delayed effects of storm sewer overflows on the oxygen conditions in rivers are also described.

Besides these specialized models, more general river water quality models have been developed which take into consideration the processes mentioned in the first part of this paper in combination with a fully dynamic hydraulic description, e.g. MIKE 11.[40,41]

Future developments

The development of faster personal computers, access to work stations and better software tools will support the future development of user-friendly 4th generation tools. At the same time more advanced descriptions of water quality parameters such as pH variation and transformation of iron and aluminium components, which are toxic to aquatic life in rivers, will be included in such modelling systems.

The intensified development of models able to describe sedimentation, consolidation and resuspension of cohesive sediments will enhance the possibilities for a more precise modelling of the effects of sediments on the aquatic life in rivers.

Water quality models are, by definition, directed towards describing *water quality parameters* used for consent setting. In future it must be anticipated that water quality models will be expanded to describe the actual effects of sewer and storm sewer overflows directly on the aquatic life and the ecological state of the rivers. A first step in this direction may be to integrate the knowledge concerning the integrated ecotoxicological effects of the different parameters such as pH, ammonia, oxygen, etc. on fish, invertebrates and other aquatic life in rivers. Other factors such as the physical dynamics of the river should also be taken into consideration including flow velocities, water depth, etc.

The final goal is to develop modelling tools which are able to describe the ecological state of rivers depending on all major forcing functions which, among others, include man's impact due to sewer and storm sewer overflows.

References

1. Streeter, H.W. and Phelps, E.B. (1925) A study of the pollution and natural purification of the Ohio river. III factors concerning the phenomena of oxidation and re-aeration, *Public Health Bulletin*, No. 146, U.S. Public Health Service.
2. Field, R. and Pitt, R.E. (1990) Urban-induced discharge impacts: U.S. Environmental Protection Agency research program review, *Wat. Sci. Tech.*, **22**: (10/11), 1–7.
3. Seager, J. and Abrahams, R.G. (1990) The impact of storm sewage discharges on the ecology of a small urban river, *Wat. Sci. Tech.*, **22**: (10/11), 147–154.
4. Danish Environmental Protection Agency (1985) Law information, *Publication No. 1*, (in Danish).
5. The Water Act, (1989) U.K.
6. Newman, P.J. (1988) Classification of surface water quality review of schemes used in EC member states, *Water Research Centre*.
7. Council of European Communities (1976) Directive 76/160/EEC concerning the quality of bathing water, *Official Journal of the European Community*, No. L31/1.
8. Council of European Communities (1991) Directive 91/271/EEC concerning treatment of domestic sewage, *Official Journal of the European Community*, No. L135/40.
9. Gromiec, M.J. (1989) Re-aeration, in: S.E. Jørgensen and M.J. Gromiec, (eds), *Mathematical submodels in water quality systems*, Elsevier, Amsterdam, 33–64.
10. O'Connor, D.J. and Di Toro, D.M. (1970) Photosynthesis and oxygen balance in streams, *J. Sanit. Eng. Div. Am. Soc. Civ. Engrs*, **96**: SA2, 547–571.
11. Alfi, S., Argaman, Y. and Shelef, G. (1972) Mathematical model for the prediction of dissolved oxygen levels in Alexander Stream, *Proc. 6th Int. Water Pollution Research*, (17) C 8/17/1–8.

12. Odum, H.T. (1956) Primary production in flowing waters, *Limnology and Oceanography*, **1**: 102–117.
13. Dobbins, W.E. (1964) BOD and oxygen relationships in streams, *J. San. Eng. Div. Am. Soc. Civ. Engrs.*, **90**: SA3, 53–78.
14. Harremoës, P. (1982) Immediate and delayed oxygen depletion in rivers, *Water Res.*, **16**: 1093–1098.
15. Hvitved-Jacobsen, T. (1982) The impact of combined sewer overflows on the dissolved oxygen concentration of a river, *Water. Res.*, **16**: 1099–1105.
16. Meinholtz, T.L., Kreutzberger, W.A., Harper, M.E. and Fay, K.J. (1979) Verification of the water quality impacts of combined sewer overflow, EPA-600/2–79–155.
17. Sill, G.C. and Thomas, R.C. (1983) Settlement and consolidation in the laboratory of steadily deposited sediment, *I.U.T.A.M. Conference on Sea-Bed Mechanics*, Newcastle.
18. Fritsch, D., Teisson, Ch. and Manoka, B. (1989) Long term simulation of suspended sediment transport. Application to the Loire Estuary, *IAHR Conference*, Ottawa, Canada.
19. Olesen, K.W. and Kjelds, J.T. (1991) Modelling of alluvial cohesive sediment transport processes, *International Symposium on the Transport of Suspended Sediments and its Mathematical Modelling*, Florence, Italy.
20. Doudouroff, P. and Skumway, D.L. (1970) Dissolved oxygen requirements of freshwater fishes, *Fd. Agric. Organ.*, **86**: 2191.
21. Davis, J.C. (1975) Minimal dissolved oxygen requirements of aquatic life with emphasis on Canadian species: a review, *J. Fish. Res.*, **32**: 2295–2332.
22. Thurston, R.V., Phillips, G.R., Russo, R.C. and Hinkins, S.M. (1981) Increased toxicity of ammonia to rainbow trout resulting from reduced concentrations of dissolved oxygen, *Can. J. Fish. Aquat. Sci.*, **38**: 983–988.
23. Thurston, R.V., Russo, R.C., Meyen, E.L. and Zajdel, R.K. (1986) Chronic toxicity of ammonia to fathead minnows, *Trans. Am. Fish. Soc.*, **115**: 196–207.
24. Simonsen, J.F. and Harremoës, P. (1978) Oxygen and pH fluctuations in rivers, *Wat. Res.*, **12**: 477–489.
25. Howard, J.R., Skirrow, G. and House, W.A. (1984) Major ion and carbonate system chemistry of a navigable freshwater canal, *Freshwater Biology*, **14**: 515–532.
26. Barker, J.P. and Schofield, C.L. (1982) Aluminium toxicity to fish in acidic waters, *Water, Air and Soil Pollution*, **18**: 289–309.
27. Munic, J.P. and Leirestad, H. (1980) Toxic effects of aluminium on the brown trout, Salmo trutta L, in: Drablos and Tollans (eds), *Ecological Impact of Acidic Precipitation*, Oslo, 84–92.
28. Garric, J., Migeon, B. and Vindimian, E. (1990) Lethal effects of draining on brown trout. A predicitive model based on field and laboratory studies, *Wat. Res.*, **24**: No. 1, 59–65.
29. Oudin, L.C. (1990) Modélisation de l'eutrophication en Loire. Modèle POPULA, *La Houille Blanche*, **4**: No. 3.
30. Bujon, G. (1988) Prévision des débits et des flux polluants transités par les réscaux d'égouts par temps de pluie. Le modèle FLUPOL, *La Houille Blanche*, No. 1.
31. Lindberg, S. and Willemoës-Jørgensen, T. (1986) MOUSE – Modelling of urban storm sewer systems, *Int. Symp. on Comparison of Urban Drainage Models with Real Catchment Data*, Dubrovnik.
32. Payne, J.A., Moyes, G.D., Hutchings, C.J. and Henderson, R.J. (1990) Development, calibration and further data requirements of the sewer flow quality model MOSQITO, *Wat. Sci. Tech.*, **22**: No. 10/11, 103–109.
33. Water Resources Engineers (1977) Computer program documentation for stream

quality model (QUAL II), U.S. Protection Agency Center for Water Quality Modelling, Athens, GA.

34. Abbott, M.B., Dahl-Madsen, K.I., Hinstrup, P.I. and Kej, A. (1975) The water quality status of the SYSTEM11–SIVA, *16th Conference* Sao Paolo.

35. Abbott, M.B., Dahl-Madsen, K.I., Hinstrup, P.I., Wium, M.P. and Verwey, A. (1975) River and estuary modelling with the SIVA system, Modelling – 75, *Int. Symp.*, San Francisco.

36. Abbott, M.B. (1989) Review of recent developments in coastal modelling, *Int. Conf. on Hydraulic and Environmental Modelling of Coastal, Estuarine and River Waters*, Bradford.

37. Crockett, C.P., Crabtree, R.W. and Cluckie, I.D. (1989) River quality models for consent setting in England and Wales, *Wat. Sci. Tech. Brighton*, **21**: 1015–1024.

38. Schaarup-Jensen, S. and Hvitved-Jacobsen, T. (1990) Dissolved oxygen stream model for combined sewer overflows, *Wat. Sci. Tech.*, **22**: No. 10/11, 137–146.

39. Crocket, C.P., Crabtree, R.W. and Markland, H.R. SPRAT – A simple river quality impact model for intermittent discharges.

40. Olesen, K.W., Havnø, K. and Malmgren-Hansen, A. (1989) A water quality modelling package for fourth generation modelling, *Proc. IAHR Congress*, Ottawa.

41. Bach, H.K., Brink, H., Olesen, K.W. and Havnø, K. Application of PC-based models in river water quality modelling.

5 Research developments of flow and water quality modelling in coastal and estuarine waters

Roger A. Falconer

Abstract

The paper describes a number of recent research developments which have been undertaken in connection with the refinement and application of a two-dimensional depth integrated numerical model for predicting tidal currents and solute concentration distributions in coastal and estuarine waters. The main refinements reported herein include:– (i) improvements in the turbulence modelling using the k-ϵ turbulence model, (ii) comparison of nested and patched modelling techniques, (iii) improvements to the representation of flooding and drying and wind stress effects, and (iv) application of a modified version of the QUICK scheme for modelling high solute gradients. These model refinements have been developed for model applications to:– (i) an idealized rectangular harbour laboratory study, (ii) nitrate level predictions in Poole Harbour, and (iii) faecal coliform levels predicted around Bridlington Sea Outfall. The resulting model refinements were shown to give improved predictions of the tidal current structure and the solute concentration distributions in all cases.

Key words: Mathematical modelling. Water quality. Tides. Coastal hydrodynamics. Turbulence. Flooding and drying. Wind Stress.

Introduction

With the growing national and international concern relating to coastal and estuarine pollution, and the constraints regarding the use of physical hydraulic models for assisting in environmental impact assessment studies, there has been a significant increase in the use of numerical models for such studies in recent years.[1] The main disadvantages in using physical models for water quality studies include:– (i) the distorted scaling of parameters, such as decay rates and mixing processes, (ii) cost, wherein such models require large and well equipped laboratories, (iii) inflexibility, in that geometric, topographic and hydrodynamic changes cannot readily be included, (iv) non-transportability, in that models built in the UK, for example, cannot be transported overseas, and (v) non-adaptability, whereby a model of the Humber Estuary, for example, cannot be used to model conditions in the Thames Estuary.

In comparison with physical models, numerical models involve modelling flow and pollutant transport processes at the prototype scale, and are generally considerably cheaper and more flexible, transportable and adaptable. On the other hand, numerical models have their disadvantages too. Apart from requiring a high level of technical expertise, numerical models involve the solution of complex differential equations governing the conservation of mass, momentum and solute transport for the flow and water quality parameters respectively. However, the true solution of the flow and solute transport processes in coastal and estuarine waters depends upon how accurately the solution of these equations, and the equations themselves, reflect the actual physical conditions in the domain. In modelling the tidal flow and solute transport processes in any coastal or estuarine basin there are still a number of uncertainties involved including, for example:– (i) the flow features, or hydrodynamics, e.g. turbulence, (ii) the physical processes, e.g. erosion and deposition of cohesive sediments, (iii) the chemical and biological processes relating to water quality parameters, e.g. decay rates for nitrates, (iv) the numerical methods, e.g. the treatment of mathematical discontinuities, and (v) the inclusion of boundary conditions, e.g. bathymetric data, tidal currents, water levels, bed roughness lengths etc.

Many of these uncertainties in current numerical models are being addressed through existing research projects, so that the predictive capability of such models can be continuously improved. Several recent research developments are outlined herein for a two-dimensional depth integrated numerical model, with particular application to modelling flow and solute transport processes in an idealized laboratory tidal basin and for various site specific studies where extensive field data have been provided.

Basic model equations

The basic hydrodynamic equations used in the model described herein were obtained by depth integration of the mass conservation (or continuity) equation and the momentum equations in the x,y co-ordinate directions in the horizontal plane.[2] The corresponding equations can be written as:–

$$\frac{\partial \zeta}{\partial t} + \frac{\partial UH}{\partial x} + \frac{\partial VH}{\partial y} = 0 \tag{1}$$

$$\frac{\partial UH}{\partial t} + \beta \left[\frac{\partial U^2 H}{\partial x} + \frac{\partial UVH}{\partial y} \right] - fVH + gH \frac{\partial \zeta}{\partial x} - \frac{C_d \, \rho_a W_x W_s}{\rho}$$

$$+ \frac{gUV_s}{C^2} - \bar{\epsilon} H \left[\frac{\partial^2 U}{\partial x^2} + \frac{\partial^2 U}{\partial y^2} \right] = 0 \tag{2}$$

$$\frac{\partial VH}{\partial t} + \beta \left[\frac{\partial UVH}{\partial x} + \frac{\partial V^2 H}{\partial y} \right] + fUH + gH \frac{\partial \zeta}{\partial y} - \frac{C_d \rho_a W_y W_s}{\rho}$$

$$+ \frac{gVV_s}{C^2} - \bar{\epsilon} H \left[\frac{\partial^2 V}{\partial x^2} + \frac{\partial^2 V}{\partial y^2} \right] = 0 \tag{3}$$

where ζ = water surface elevation above datum, t = time, U,V = depth average velocity components in x,y directions, H = total depth of flow, β = momentum correction factor for non–uniform vertical velocity profile, f = Coriolis parameter, g = gravitational acceleration, C_d = air-water resistance coefficient, ρ_a = density of air, W_x, W_y = wind velocity components in x,y directions, W_s = wind speed, ρ = density of water, V_s = fluid speed, C = de Chezy roughness coefficient and $\bar{\epsilon}$ = depth average eddy viscosity.

In practical model studies, and in the absence of field data, the momentum correction factor β is either set to unity or a vertical velocity profile is assumed, such as a logarithmic velocity profile.[2] For the Chezy value either a constant value can be included directly, or C can be evaluated from the Manning equation or, for enhanced accuracy, from the Colebrook-White equation.[3] Likewise, as for values of β, the depth average eddy viscosity can be obtained either from field data or calculated analytically provided that bed generated turbulence dominates over free shear turbulence, and assuming a logarithmic vertical velocity profile.[4] For the solute transport process modelling, the depth integrated form of the advective-diffusion equation can be written in the following form:–

$$\frac{\partial H\phi}{\partial t} + \frac{\partial UH\phi}{\partial x} + \frac{\partial VH\phi}{\partial y} - \frac{\partial}{\partial x}\left[HD_{xx}\frac{\partial \phi}{\partial x} + HD_{xy}\frac{\partial \phi}{\partial y} \right]$$

$$- \frac{\partial}{\partial y}\left[HD_{yx}\frac{\partial \phi}{\partial x} + HD_{yy}\frac{\partial \phi}{\partial y} \right] - H\left(S_L + S_B + S_k \right) = 0 \qquad (4)$$

where ϕ = depth average solute concentration level, D_{xx}, D_{xy}, D_{yx}, D_{yy} = depth average longitudinal dispersion and turbulent diffusion coefficients, S_L = direct and diffuse loading rate, S_B = boundary loading rate and S_k = total kinetic transformation rate.

Numerical solution

The governing hydrodynamic and solute transport equations given in Eqs. (1)–(4) are generally solved using either finite difference or finite element techniques. For reasons of computational efficiency and for minimal artificial diffusion, the finite difference solution technique has been much more widely used for such model studies and the research developments reported herein refer to a finite difference model.

The model developed is based on an alternating direction implicit finite difference scheme, with the advective accelerations in the momentum equations and the turbulent diffusion terms being centred by iteration. The scheme adopted is unconditionally stable, although for accuracy requirements the Courant number (C_r) is limited accordingly:–

$$C_r = \frac{\Delta t}{\Delta x}\sqrt{(gH)} \leq 8 \qquad (5)$$

where Δt = time step size and Δx = grid size. Further details of the general numerical model are given in Falconer.[5]

Research developments

The main recent research developments undertaken by the author relating to water quality modelling in coastal and estuarine waters include:– (i) improvements in the turbulence modelling, (ii) comparison of nested and patched modelling techniques, (iii) improvements in the representation of flooding and drying of tidal flood plains, and (iv) application of a modified version of the QUICK scheme for modelling high solute gradients. These developments have been made for applications to:– (i) an idealized rectangular harbour laboratory study, (ii) nitrate level predictions in Poole Harbour and (iii) faecal coliform levels predicted around Bridlington Sea Outfall. These research developments are summarised accordingly:–

(i) Turbulence modelling developments

In modelling flow and water quality processes in coastal and estuarine waters, the occurrence of tidal eddies is important to the planner and coastal engineer in terms of their impact on sediment deposition, pollutant mixing and the siting of sea outfalls etc. Such time dependent tidal eddies are complex hydrodynamic flow processes which are not easy to model numerically, and laboratory model scale predictions can frequently lead to erroneous results, particularly when Froudian law distorted models are used.

The results of an analytical study of the depth average vorticity transport equation describing tidal eddies[6] has shown that the advective accelerations and lateral turbulent shear stresses are important, and that bed slopes can be significant in generating such eddies through the use of a quadratic friction law. Hence, it is essential to ensure that complex bed topographies are modelled accurately in practical numerical model simulations.

In attempting to model tidal eddies accurately, two specific turbulence models were considered and compared. The first involved the application of a zero-equation mixing length model, including an empirical representation for free shear layer turbulence, and the more complex two-equation unsteady k–ϵ turbulence model. The numerical representation of the time dependent k–ϵ differential equations was found to exhibit stability problems using a second order accurate implicit finite difference scheme and an intolerable level of artificial diffusion occurred when a first order accurate scheme was used.

In testing the different turbulence models, the mathematical model was applied to three studies, including:– (i) tide induced circulation in a rectangular laboratory model harbour, (ii) tidal eddy circulation in the lee of Rattray Island, Australia, and (iii) tidal circulation in Port Talbot Harbour, UK.

In the first of these applications detailed laboratory measurements were made of the depth mean velocity profiles along the central axes AOC and DOB as shown in Figure 1. The velocity measurements were taken at 10s intervals, at one minute either side of mean water level, for sinusoidal model tides of period 708s. The model harbour was located at the rear end of a tidal basin, as shown in Figure 2, with an oscillating overflow weir being driven by computer and generating tides of variable amplitude and period. The model harbours considered were distorted scaled laboratory models of an idealized prototype square harbour, of length 432m × 432m, with a resultant plan surface area of 18.7ha. The z horizontal scale ratio was varied from 1:200 to 1:400 and with the vertical scale varying from 1:12 to 1:3.125. Full details of the laboratory configuration and experimental procedure are given in Falconer and Yu.[7]

The model grid spacing was typically 60mm and the time step was chosen such that the maximum Courant number was approximately 8. Typical comparisons of predicted and measured velocity distributions, both for the k-ε and the mixing length turbulence models, are shown in Figures 3 and 4, with further details being given in Falconer and Li.[8]

The main findings from this research programme can be summarized as follows:–

i) The lateral turbulent shear stresses and the advective accelerations were found to be extremely important in modelling tidal eddies.

ii) The influence of the bathymetry on the structure and strength of the eddies was also found to be particularly important.

iii) The use of a quadratic friction law was shown to give rise to components of both vorticity generation and dissipation.

iv) The choice of the grid size was found to be critical in accurately reproducing tidal eddies.

v) In an analytical and numerical model analysis of the influence of the various terms of the equations of motion on the hydro-dynamics of tidal eddies, differences in the numerical model predictions of tidal eddies in prototype rectangular harbours and their respective physical models have highlighted questions as to the suitability of using distorted physical models for such studies.

vi) The turbulent shear stress and the no-slip boundary condition were shown to have a marked effect on the tidal eddy structure near a wall, both in the prototype and physical model simulations.

vii) The unsteady k-ε turbulence equations were found to become unstable using traditional finite difference schemes and higher order accurate schemes, such as QUICK and EXQUISITE, had to be adopted.

viii) The k-ε turbulence model solutions gave markedly different eddy viscosity distributions from those predicted using simple mixing length models. Although the disparate eddy viscosity distributions showed little difference in the velocity field predictions, the effect was much more pronounced on the turbulent diffusion of conservative and non-conservative solutes.

ix) Finally, the refined unsteady turbulence model showed encouraging agreement between the laboratory results and the numerical model predictions.

(ii) Nested and patched models

A recent research programme has been undertaken to develop, refine and apply two different types of combined coarse and fine grid numerical models, with such numerical models being increasingly used to obtain higher resolution of flow and water quality parameter distributions in regions of particular interest.[1] For example, a fine grid model may be used to obtain a detailed prediction of the velocity and solute distributions within a small harbour, whereas a coarse grid model can be used outside the harbour where less accuracy may be required. Furthermore, nested models are increasingly being used for hydraulic model studies where open boundary data are sparse, or nonexistent, with the coarse grid model providing the open boundary conditions in regions of interest.

The two main models considered involved a nested (non-dynamically linked) and a patched (dynamically linked) model, with there being advantages and disadvantages with both schemes. In particular, emphasis was also focussed on fully including the advective accelerations at the interface between the fine and coarse grid boundaries in the patched model, thereby allowing eddies and fine grid flow features to be advected out into the coarse grid domain. The numerical models were again applied to the laboratory harbour model studies highlighted previously, with the depth average velocities being compared for the predicted and measured velocities across the central axes.

For the nested model investigations it was found that in the coarse grid domain spurious negative velocities were predicted along the harbour entrance streamline, see Figure 5. This anomaly was first thought to be due to the inadequate resolution of the high velocity gradients in the region of the harbour entrance. However, although the use of higher order accurate difference schemes for the treatment of the advective accelerations was found to reduce the spurious velocities, the negative velocities still persisted throughout most of the flood tide. Furthermore, the nested model also predicted an ebb tide jet orientation on leaving the harbour which was normal to the plane of the harbour entrance, see Figure 6.

For the patched model investigations, the numerically predicted velocity fields were much more consistent with laboratory measurements and observations. No spurious negative velocities were predicted in the coarse grid region during the flood tide and the exhaust jet orientation agreed closely with the laboratory model observations. Furthermore, for the patched model, the exhaust jet was predicted to generate free shear eddies just outside the harbour entrance and these eddies were then advected from the fine grid to the coarse grid and into the forebay of the tidal tank, see Figure 7. Again these predicted eddies were in close agreement with the laboratory model results, although they were not predicted in the nested model simulations.

Finally, comparisons were made of the predicted and measured velocity distributions across the central axes for first order and third order upwind difference representations of the advective acceleration terms in the momentum equations. As can be seen from the results shown in Figure 8, the third order upwind differencing treatment of the advective accelerations gave closer agreement with the experimental results[9] than the original scheme, although there were some noticeable disparities between the results for low velocities of less than about $5\,\mathrm{mms}^{-1}$. However, the velocities were measured by tracking fishing floats and it was difficult to track these floats at low velocities. In particular, for the flood tide velocity profile along the AOC axis it can be seen that the third order upwind differencing scheme has predicted the measured peak jet velocity closely, whereas the first order difference scheme significantly under predicts this velocity. Other discrepancies between the measured and predicted velocities were predominantly thought to be due to the three-dimensional nature of the velocity field – particularly associated with secondary currents – and the simplicity of the velocity measuring technique. With the recent acquisition of a laser doppler anemometer, more precise laboratory measurements are now being taken.

(iii) Flooding, drying and wind stress effects

In modelling numerically the flooding and drying of shallow reaches throughout the tidal cycle, considerable numerical problems can arise as a result of the discretized representation of this hydrodynamic process, which generally varies in a smooth manner. Previous studies undertaken by Falconer[10,11] on modelling the flooding and drying process in Holes Bay and Poole Harbour, Dorset, showed that the modifications to a previous scheme described by Leendertse and Gritton[12] for modelling flooding and drying gave numerically stable and, on the whole, encouraging predictions of the tidal velocities in both coastal basins. However, for the latter study, the numerically predicted water elevations gave greater phase lags across the harbour than those measured in the field, and larger velocities were often observed immediately adjacent to a dry grid square. Although this scheme generally appeared to give satisfactory predictions of the tidal flows for both studies, the same scheme exhibited oscillations in the water level field when applied to the Humber Estuary.[13] At the time this instability was attributed to the much larger grid size, and hence the greater discretization, for the Humber Estuary and a new scheme was developed and outlined accordingly for this study. The revised scheme documented by Falconer and Owens[13] has subsequently also exhibited numerical problems in some practical applications, particularly where the flood plain changes abruptly from a nearly horizontal bed to a sloping beach. In a more recent study[14] extensive further tests have been undertaken on various representations of the flooding and drying process for a range of complex idealised estuarine bathymetries. These tests have highlighted a number of deficiencies in the previous schemes and a new scheme has been developed for modelling this complex hydrodynamic process. Apart from proving to be more accurate and robust than the previous schemes, the new scheme has also compared favourably with other schemes – such as that outlined by Stelling et al[15] – when predicted results are compared with field data for Poole Harbour in Dorset, see Figures 9 and 10 respectively.

In refining the flooding and drying process another refinement was also undertaken which involved including a more complete representation of the effects of a surface wind stress. In most two-dimensional depth average tidal numerical models reported in the literature to-date the effects of a surface wind stress have been included in the model solely in terms of an additional shear stress term at the free surface, for example see Falconer.[11] However, in shallow water coastal and estuarine flows a surface wind stress will affect the predicted water elevations and depth average velocities in a two-dimensional hydrodynamic model by two processes. Firstly, the wind action on the surface will affect the longitudinal pressure gradient, thereby changing the water elevation and, indirectly, the depth average velocity. Secondly, a wind stress will change the form of the vertical velocity profile which, in turn, will change the momentum correction factor β and the form of the advective accelerations. Where extensive field data of the vertical velocity profile is available for model calibration with a wind stress, then the value of β can be evaluated directly from the field data and the advective accelerations need not be modified accordingly, see Falconer and Owens.[13] However, in most practical model applications, sufficient data regarding the vertical velocity profile for a wide range of wind velocities are not available for model calibration and approximations therefore have to be made as to the form of the vertical velocity profile. In general the form of the vertical velocity profile to-date has either been ignored, e.g. Stelling et al,[15] or approximated by a logarithmic or seventh power law velocity distribution, e.g. Falconer.[11] However, from analytical solutions and field or laboratory data it has been established that wind generated velocity profiles in semi-enclosed coastal and estuarine basins vary considerably over the depth,[16] with the vertical velocity profile generally being more accurately represented using a second order parabola of the following form, given for the x–direction as:–[14]

$$u = \left[\frac{3C_x}{4} - \frac{3U}{2} \right] \left[\left(\frac{z}{H} \right)^2 - 1 \right] + C_x \left[\left(\frac{z}{H} \right) + 1 \right] \tag{6}$$

where u = velocity at elevation z, with $z = 0$ at the surface and $z = -H$ at the bed, and $C_x = \tau_{sx} H / \rho \bar{\epsilon}$, where τ_{sx} = surface shear stress component in the x-direction.

The corresponding form of the advective accelerations for Eq.(6) has been established along similar lines to that outlined by Koutitas and Gousidou-Koutita[16] except that the advective accelerations have been expressed in their pure differential form giving a more accurate physical representation and, in particular, allowing a fully conservative finite difference representation. The modification required to the advective accelerations for the x-direction momentum equation (i.e. Eq. 2) can be shown to be of the form:–[14]

$$\textit{Advective Acceleration} = \frac{\partial}{\partial x} \left[\frac{C_x H U}{20} + \frac{C_x^2 H}{120} \right]$$

$$+ \frac{\partial}{\partial y} \left[\frac{C_y H U}{40} + \frac{C_x H V}{40} + \frac{C_x C_y H}{120} \right] \tag{7}$$

with a similar modified term being derived for the y-direction.

The corresponding modified numerical model has again been applied to Poole Harbour for mild wind speeds, with the comparability with field measured data being more encouraging than for inclusion of a simple seventh power law. To illustrate the predicted wind effect on the velocity field, the near surface and near bed tidal currents are shown in Figures 11 and 12 respectively for a south westerly wind of $3ms^{-1}$. From these graphs it can be seen that for the near surface velocity field predictions two counter-rotating eddies are predicted immediately inside the harbour entrance, both to the south west and north east of the entrance, whereas for the near bed velocity field predictions the velocities are considerably reduced and no distinct flow features can be identified. Comparisons have been made at a number of sites across the basin of predicted and measured velocities at 1m below the surface and above the bed. In general the corresponding results compared more favourably than for the inclusion of a seventh power law velocity profile.[14]

(iv) Application of the QUICK scheme for modelling high solute gradients

With increasing use being made of numerical models for water quality studies relating to coastal and estuarine waters, increasing emphasis has been focussed on the numerical treatment of the advective terms of the solute transport equation. When a time centred second order accurate representation is used for these terms, then pronounced grid scale oscillations are generally observed in regions of high concentration gradients. As time progresses these oscillations can propagate across the domain – particularly when the physical diffusion is relatively small – and the mass conservation solution can become meaningless physically as a result of the occurrence of negative concentrations.[17]

In overcoming or reducing these problems associated with a second order accurate central difference representation of the advection terms, increasing use has been made of higher order accurate schemes. For this purpose the QUICK difference scheme[18] has become increasingly attractive due to its computational efficiency and simplicity. The QUICK scheme is based on assuming quadratic interpolation – rather than linear – interpolation between the grid points and was first proposed by Leonard[18] for steady flows. When the scheme is applied to unsteady tidal flows, several different finite difference forms exist.[19] These schemes include:– (i) the explicit forward QUICK scheme, which has a severe stability constraint, (ii) the fully time centred implicit QUICK scheme, which is unconditionally stable, (iii) the backward implicit QUICK scheme, which is conditionally stable and computationally efficient, and (iv) the semi-implicit QUICK scheme, which is conditionally stable but diffusive. These modified schemes have been applied to, and compared for, three idealized one-dimensional test cases with no diffusion, including advection of:– (i) a plug source, (ii) a

Gaussian concentration distribution, and (iii) a sharp front concentration gradient. The results of these tests showed that the QUICK schemes (iii) and (v) were more diffusive than schemes (ii) and (iv), whereas schemes (ii) and (iv) may exhibit some overshoot and undershoot for steep gradients – albeit that this effect is small in comparison with that obtained using a central difference second order representation of the advection terms.

Having compared various representations of the QUICK scheme for one-dimensional test cases, these schemes were than applied to Poole Harbour. The original study was undertaken for Wessex Water, whose main interest was to ascertain the influence of the present levels of nitrogen input from Poole Sewage Works on the corresponding concentrations across the whole basin, particularly in view of the growth of the green seaweeds Ulva and Enteromorpha in the region of the Main Channel and Wych Channel, to the north east of Brownsea Island. The nitrogen inputs of particular interest in this study were the nitrate concentrations resulting from inputs of both total oxidised nitrogen and ammoniacal nitrogen from two river inputs and three sewage works. The corresponding predictions of the nitrate levels obtained using the semi-implicit QUICK scheme and the second order central difference scheme[11] are shown in Figures 13 and 14 respectively. Although the difference between the contour levels for both plots is not significant, the corresponding numerical values often showed marked differences. In particular, in the region of interest just to the north east of Brownsea Island – contained within the $0.1mg\ l^{-1}$ contour – the nitrate levels using the QUICK scheme varied smoothly and were in closer agreement with the measured field data. In contrast, the central difference scheme, although not diffusive, produced a wave type spatial nitrate distribution within this region, with concentration levels dropping to as low as $-0.7mg\ l^{-1}$. Clearly, although such values might give rise to conservative and non-diffusive overall results, they become meaningless once decay rates and pollutant interactions are considered.

Conclusions

With an increasing use being made of numerical hydraulic models for predicting flow and water quality processes in coastal and estuarine waters, there has correspondingly been an increasing need for research developments to be made in connection with such models. Although extensive research programmes are currently underway on a wide range of studies relating to refinements and developments to numerical models, details are given herein of some recent research programmes undertaken by the Computational Hydraulics and Environmental Modelling Research Group at the University of Bradford.

The research developments reported in the paper apply to a two-dimensional depth integrated flow and solute transport model and include:–

(i) The Inclusion of a Refined Turbulence Model of the k-ε Type. The results showed that although the turbulence model did not have a significant effect on the gross tidal flow structure within a rectangular harbour, fine flow details were predicted more accurately and, in particular, the turbulent diffusion coefficient was more precise.

(ii) A Comparison of Nested and Patched Modelling Techniques. The results showed that for certain types of flow fields nested modelling techniques may give inaccurate and sometimes spurious velocity field predictions. On the other hand dynamically linked, or patched models do not exhibit these inaccuracies or spurious results. Both schemes produced more accurate results when a third order accurate scheme was used to model the advective accelerations.

(iii) The Enhancement of Flooding and Drying and Wind Stress Effects. A flooding and drying algorithm developed by the author has been refined to give improved accuracy of this complex hydrodynamic process and the representation of a surface wind stress has been improved by modifying the form of the advective accelerations as a result of assuming that a second order parabola more closely represents the vertical velocity field distribution for wind induced flows. The results for both of these refinements showed improved predictions of the hydrodynamic field data provided for tidal currents in Poole Harbour, Dorset, with the flooding and drying scheme showing enhanced accuracy in comparison with some other models reported in the literature.

(iv) The Application of a Modified Form of the QUICK Scheme. For modelling high solute gradients the commonly used second order accurate central difference scheme has a number of disadvantages and the QUICK scheme has been modified to model this commonly occurring phenomenon. The resulting scheme has been applied to idealized test cases and nitrate level predictions in Poole Harbour, with both sets of results showing improved solute level predictions in comparison with second order accurate schemes. Further research is continuously being undertaken by the Research Group at the University of Bradford on the development, verification and refinement of numerical models for flow and water quality modelling in coastal and estuarine waters.

Acknowledgments

Several of the research projects reported herein have been funded by the Science and Engineering Research Council, IBM UK Ltd, Wessex Water plc and Sir William Halcrow & Partners. Data for the studies relating to Poole Harbour have been provided by Wessex Water plc, Poole Harbour Commissioners and Hydraulics Research Ltd. The Author is grateful to these organizations for their support.

The Author is also grateful to the following research assistants and students who collaborated on the various projects:– Dr P.H. Owens, Dr G. Li, Mr Y. Chen and Mr R.D. Alstead.

References

1. Falconer, R.A. (1992) Flow and Water Quality Modelling in Coastal and Inland Waters, *Journal of Hydraulic Research*, (in press).
2. Falconer, R.A. (1991) Review of Modelling of Flow and Pollutant Transport Processes in Hydraulic Basins, *Proceedings of First International Conference on Water Pollution*, Computational Mechanics Institute, Southampton, 3–5 September, pp.1–21.
3. Henderson, F.M. (1966) *Open Channel Flow*, Collier-Macmillan Publishing Co. Inc., New York, pp.522.
4. Fischer, H.B. (1976) Mixing and Dispersion in Estuaries, *Annual Review of Fluid Mechanics*, **8**, pp.107–133.
5. Falconer, R.A. (1986) A Two-Dimensional Mathematical Model Study of the Nitrate Levels in an Inland Natural Basin, *Proceedings of the International Conference on Water Quality Modelling in the Inland Natural Environment*, BHRA Fluid Engineering, Bournemouth, Paper J1, June, pp.325–344.
6. Falconer, R.A. and Mardapitta-Hadjipandeli, L. (1987) Bathymetric and Shear Stress Effects on an Island's Wake : A Computational Model, *Coastal Engineering*, Elsevier Scientific Publications, **11**, No.1, March, pp.57-86.
7. Falconer, R.A. and Yu, G.P. (1991) The Effects of Depth, Bed Slope and Scaling on Tidal Currents and Exchange in a Laboratory Model Harbour, *Proceedings of the Institution of Civil Engineers, Part 2, Research and Theory*, September, pp.561–576.
8. Falconer, R.A. and Li, G. (1989) Numerical Modelling of Tidal Eddies in Narrow Entranced Coastal Basins and Estuaries, SERC Report GR/E/42655, December, pp.1–25.
9. Nece, R.A. (1990) Tidal Current Measurements in a Laboratory Harbour, *Journal of Waterway, Port, Coastal and Ocean Engineering*, ASCE, submitted for publication.
10. Falconer, R.A. (1984) A Mathematical Model Study of the Flushing Characteristics of a Shallow Tidal Bay, *Proceedings of the Institution of Civil Engineers*, Part 2, **77**, September, pp.311–332.
11. Falconer, R.A. (1986) A Water Quality Simulation Study of a Natural Harbour,

Journal of the Waterway, Port, Coastal and Ocean Engineering Division, ASCE, **112**, No.1, January, pp.15–34.

12. Leendertse, J.J. and Gritton, E.C. (1971) A Water Quality Simulation Model for Well-Mixed Estuaries and Coastal Seas : Volume 2, Computation Procedures, The Rand Corporation, Santa Monica, Report No.R–708–NYC, July, pp.1–53.

13. Falconer, R.A. and Owens, P.H. (1987) Numerical Simulation of Flooding and Drying in a Depth-Averaged Tidal Flow Model, *Proceedings of the Institution of Civil Engineers*, Part 2, **83**, March, pp.161–180.

14. Falconer, R.A. and Chen, Y. (1991) An Improved Representation of Flooding and Drying and Wind Stress Effects in a 2-D Tidal Numerical Model, *Proceedings of the Institution of Civil Engineers, Part 2, Research and Theory*, **91**, December, pp.659–678.

15. Stelling, G.S., Wiersma, A.K. and Willemse, J.B.T.M. (1986) Practical Aspects of Accurate Tidal Computations, *Journal of Hydraulic Engineering*, ASCE, **112**, No.9, September, pp.802–817.

16. Koutitas, C. and Gousidou-Koutita, M. (1986) A Comparative Study of Three Mathematical Models for Wind Generated Circulation in Coastal Areas, *Coastal Engineering*, **10**, pp.137–138.

17. Leendertse, J.J. (1970) A Water Quality Simulation Model for Well Mixed Estuaries and Coastal Seas: Volume 1, Principles of Computation, Rand Corporation Report No. RM–6230–RC, February, pp.1–71.

18. Leonard, B.P. (1979) A Stable and Accurate Convective Modelling Procedure Based on Quadratic Upstream Interpolation, Computational Methods for Applied Mechanical Engineering, **19**, pp.59–98.

19. Chen, Y. and Falconer, R.A. (1992) Advective-Diffusion Modelling Using the QUICK Difference Scheme, submitted for publication to *International Journal of Numerical Methods in Fluids*.

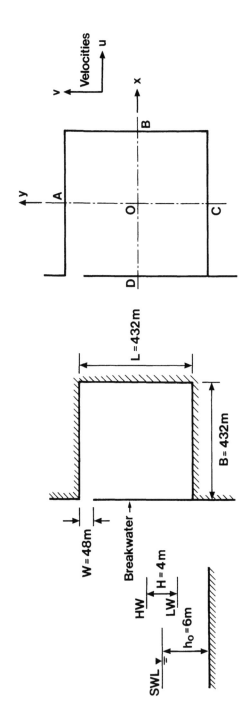

Figure 1 Schematic illustration of idealized prototype harbour, showing tidal range, mean depth and axes

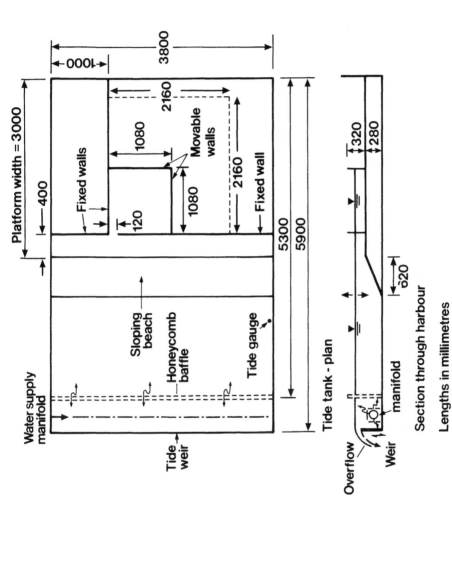

Figure 2 Illustration of the tidal basin showing the numerical model open boundaries

97

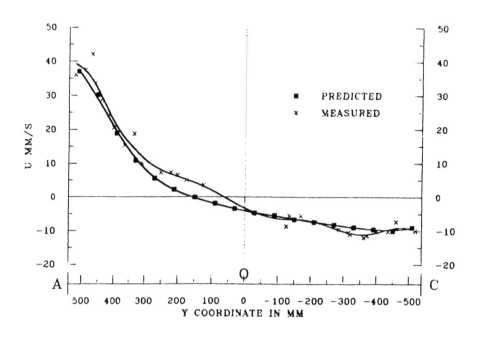

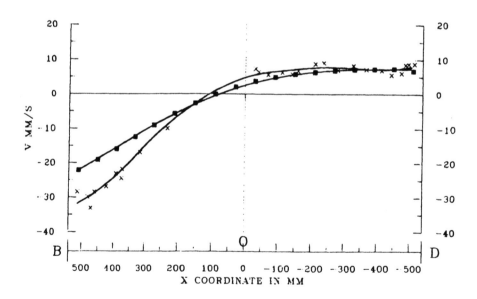

Figure 3 **Measured and predicted velocity profiles along the axes at mean water level flood tide using the k-ε model**

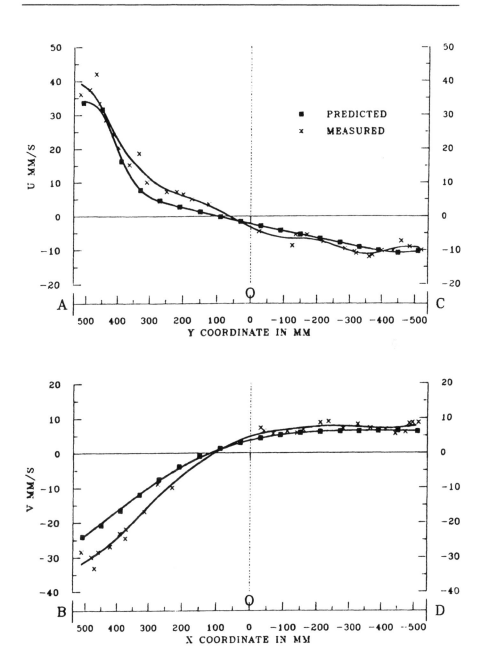

Figure 4   Measured and predicted velocity profiles along the axes at mean water level flood tide using the modified mixing length model

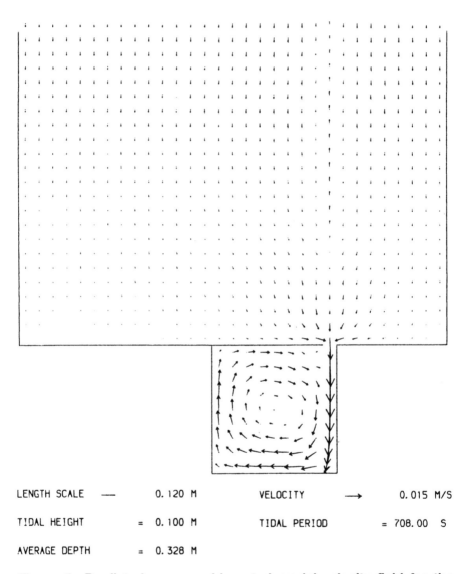

TIDAL VELOCITIES IN A HARBOUR

TIME= 885. 60 S

LENGTH SCALE — 0. 120 M	VELOCITY ⟶ 0. 015 M/S	
TIDAL HEIGHT = 0. 100 M	TIDAL PERIOD = 708. 00 S	
AVERAGE DEPTH = 0. 328 M		

Figure 5   Predicted coarse grid nested model velocity field for the flood tide showing the spurious velocities beyond the model harbour entrance

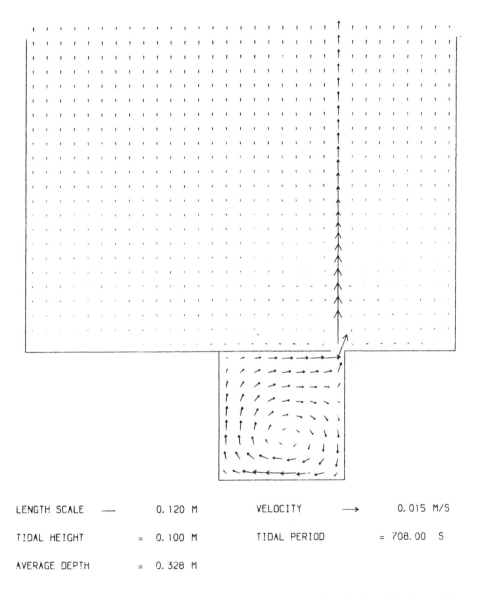

LENGTH SCALE — 0.120 M   VELOCITY → 0.015 M/S

TIDAL HEIGHT = 0.100 M   TIDAL PERIOD = 708.00 S

AVERAGE DEPTH = 0.328 M

Figure 6   Predicted coarse grid nested model velocity field for the ebb tide showing the incorrect jet orientation

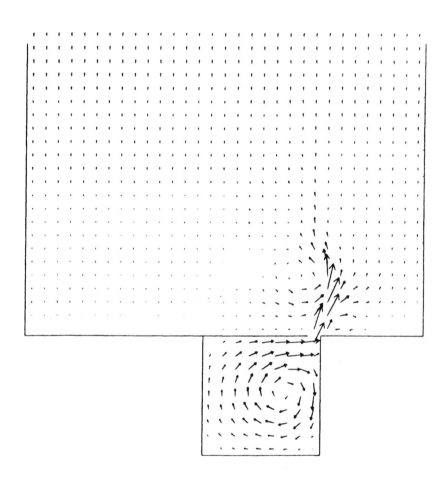

TIDAL VELOCITIES IN A HARBOUR

TIME= 531.10 S

LENGTH SCALE — 0.120 M	VELOCITY ⟶ 0.015 M/S
TIDAL HEIGHT = 0.100 M	TIDAL PERIOD = 708.00 S
AVERAGE DEPTH = 0.328 M	

Figure 7   Predicted coarse grid patched model velocity field for the ebb tide showing the correct jet orientation

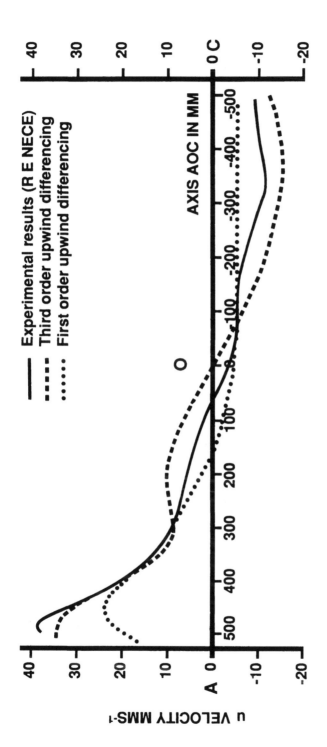

Figure 8   Comparison of predicted and measured velocity distributions across the model

103

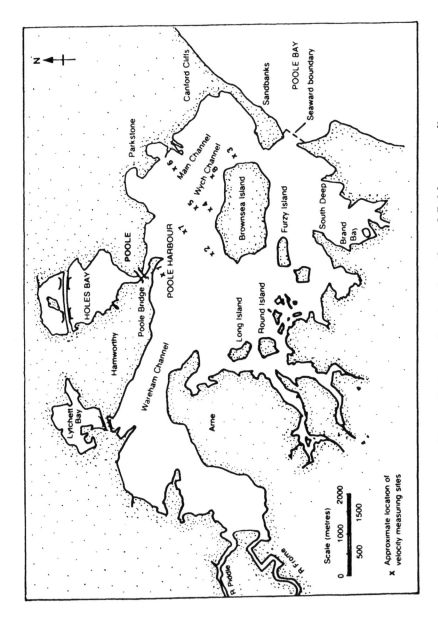

Figure 9 Map of Poole Harbour showing location of field survey sites

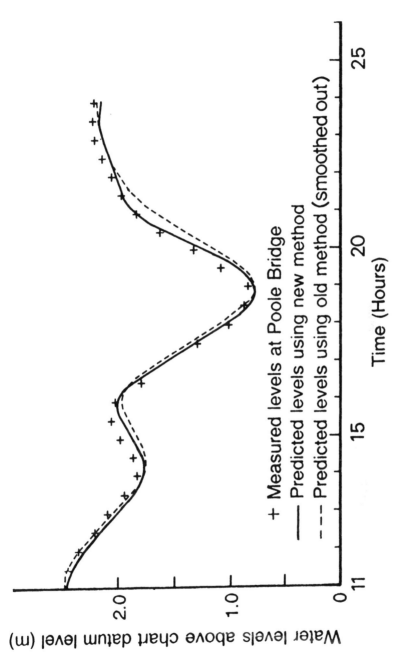

Figure 10 Comparison of predicted water elevations at Poole Bridge using the old and new methods for flooding and drying

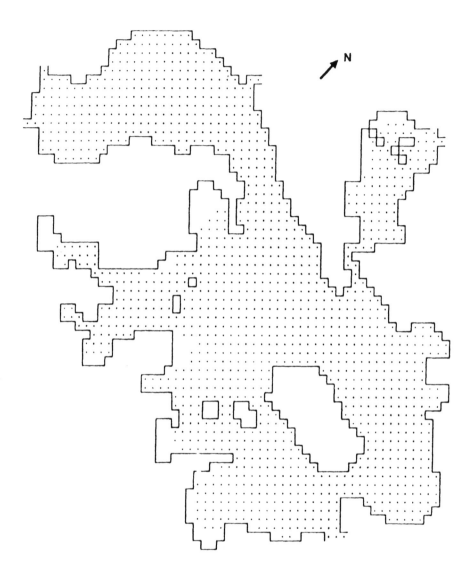

TIME = 21.00 HOURS   Near Bottom Velocity
GEOMETRIC SCALE: □   150m x 150m
VELOCITY SCALE: →   1.00 m/s

Figure 11 Predicted flood tide near surface velocities in Poole Harbour

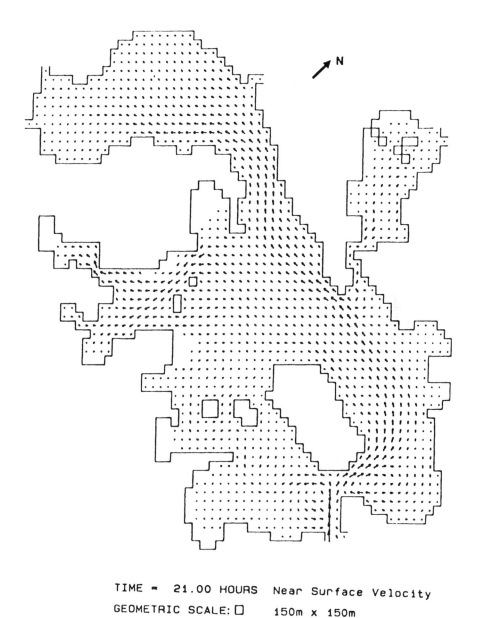

TIME = 21.00 HOURS   Near Surface Velocity
GEOMETRIC SCALE: ☐   150m x 150m
VELOCITY SCALE: →   1.00 m/s

Figure 12   Predicted near bed velocities in Poole Harbour

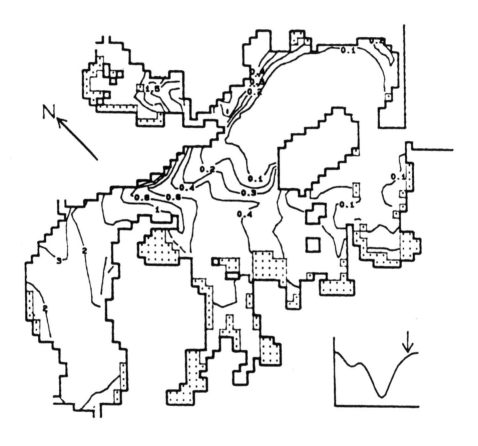

Figure 13 Nitrate level predictions at high tide (in mgl⁻¹) using the QUICK difference scheme

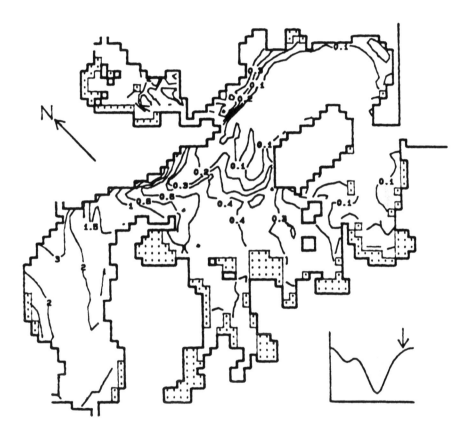

Figure 14 Nitrate level predictions at high tide (in mgl⁻¹) using the central difference scheme

6 Hydrodynamic and physical considerations for water quality modelling

Mrs J.M. Maskell

Abstract

Water quality in coastal and estuarine waters is governed by the interaction of a number of hydrodynamic processes. The capability of a model to simulate and predict water quality is dependent on its ability to simulate all the relevant hydrodynamic processes. This paper describes some of the major processes which need to be considered when choosing the type of model to use for a particular application.

Key words: Gravitational circulation; Hydrodynamics; Particulate BOD; Mud Transport; Stratification; Water quality.

Introduction

Pollutants discharged into estuaries and coastal waters are dispersed not only by the action of tides but also to a greater or lesser extent by the presence of other hydrodynamic and physical phenomena. One dimensional and two dimensional-in-plan mathematical models are routinely used to predict water quality in estuaries and coastal waters and when used appropriately are very useful tools. However, a model can only simulate water quality if it is capable of reproducing the relevant hydrodynamic and physical processes which occur in the water body or which may be expected to occur as a result of proposed engineering works. There are a large number of cases where the use of depth-averaged models is, for one reason or another, inappropriate.

The ultimate model would be three dimensional, with an infinitely variable grid to resolve detail through the water column as well as in plan, and capable of including the effect of wind field and its modification by land topography, transient phenomena such as storms and the longer term changes in bathymetry. However, in practice it is often necessary to use the simplest model appropriate to a particular application. The art of the modeller is to choose the correct model and it requires experience and skill to decide which processes may be ignored or simplified. The omission of relevant processes and the resultant compensation by the use of additional empirical coefficients during calibration results in a model which cannot be predictive.

This paper describes various hydrodynamic and physical phenomena which can have significant implications with regard to water quality and the consequences for the water quality modeller.

Gravitational circulation

Gravitational circulation takes the form of a two layer exchange between buoyant river water and sea water. The combination of a longitudinal density gradient and the raised mean tidal level along the estuary, resulting from the bathymetry and the effect of the density difference, combine to generate a residual landward flow near the bed with a residual seaward flow at the surface. The inflowing seawater upwells, mixes and progressively dilutes the brackish surface water (Fig.1). The strength of the circulation varies with the magnitude of the product of the water depth and the longitudinal density near the bed. It is reduced by vertical mixing, which is heavily damped in stratified flows, and by energy dissipation at the bed, which is increased by the occurrence of high tidal velocities in the lower layers. The gravitational flow is often many times larger than the fresh water discharge and lags behind changes in freshwater flow.

The strength of the circulation may vary according to the degree of stratification but it is not dependent on the existence of vertical density stratification. Many deep estuaries and coastal waters with weak and negligible stratification, such as the Thames estuary, have strong gravitational circulations.

The limit of the circulation is at the point where the effect of the longitudinal density gradient is balanced by the mean tide slope. A turbidity maximum may occur at this point as suspended mud flocs settle into the lower layer and are transported landward by the gravitational circulation.

The gravitational circulation can only be simulated in a mathematical model that takes into account the vertical structure of the flow. Depth averaged models are inadequate because they provide no information about the vertical structure of the velocity, density and concentration profiles and cannot simulate the turbidity maximum. Various methods exist by which the effects of gravitational circulation are 'approximated' in depth-averaged models. These include increasing the coefficient of effective longitudinal dispersion and assuming different bed roughnesses on the flood and ebb phases of the tide to change the shape of the theoretical vertical velocity profile. All are somewhat empirical and unsatisfactory since they do not permit prediction of changes in gravitational circulation resulting from changes in freshwater flow or as a result of major engineering works.

Stratification

Vertical mixing is caused by turbulent eddies which generate internal stresses in the flow by exchanging momentum from fast flowing layers to slower flowing layers and vice versa. Turbulence also mixes salt, heat, sediment and pollutants through the water column. The presence of stable stratification, that is density increasing downwards, caused by vertical variations in salinity or temperature damps the vertical turbulent eddies and reduces vertical mixing drastically. The damping effect is significant even at relatively low levels of stratification.

The extent of stratification and the depth of the halocline depends on the combination of freshwater flow and tidal range. In order to simulate the effects of stratification it is necessary to model the vertical structure of the flow and the damping of vertical mixing.

HR have developed a suite of two-dimensional-in-depth and three-dimensional models using analytical mixing length functions derived from the theories of Prandtl, Rosby and Montgomery and Ellison.[1] The models have been used successfully in a number of studies without any need to change the empirical coefficients. Calibration is carried out by use of a single scaling factor to define the amount of longitudinal mixing in the layers.

The effect of stratification on water quality depends on the particular conditions applying. In the Tees estuary, which is highly stratified, the majority of the polluting load is discharged into the brackish surface layers and density effects contain the effluent in the surface waters thereby reducing the potential dilution. In the lower and middle estuary relatively clean sea water is overlain by poor quality brackish water which on occasions is almost devoid of oxygen.

Results from the HR model POLLFLOW-2DV (Fig.2) show the ability of such a model to reproduce the different characteristics of the lower and upper estuary. At Middlesbrough Dock the incoming sea water in the bed layer is relatively clean whilst the effect of non-saline effluent inputs is clearly seen in the surface layer. However, at Victoria Bridge, which is upstream of the limit of stratification under low flow conditions, the water column is vertically well mixed and the temporal variation in BOD is the result of the advection of poor quality water from downstream on the flood tide.

In Hong Kong coastal waters variations in the freshwater discharge into the Pearl estuary give rise to significant seasonal variations in water movements. During the wet summer months the large Pearl River discharge results in saline stratification which effectively traps effluents discharged near the bed and prevents the downward transport of dissolved oxygen thus reaeration at the water surface has no impact on the lower layers of the water column. As a result, during the summer, water quality in the lower layer is worse than that in the surface layers and this must be simulated by using a layered model. During the drier winter months the coastal waters of Hong Kong are generally well mixed through the vertical, and short term simulations of water quality in such conditions can be undertaken using two-dimensional in plan models. However longer term simulations of nutrient balance require adequate representation of the build up of stratification, the large nutrient loadings arising from the increased river flows and the effect of gravitational circulation. Such simulations necessitate the use of three-dimensional models of flow and water quality.

Wind effects

Local winds generate a stress on the water surface which results in a surface current and a subsurface return current. Short term wind effects on buoyant effluent plumes can often be adequately represented by superposing the wind drift on top of the flow field simulated for calm conditions. However the effect of persistent winds on water quality throughout the water column requires a more rigorous examination of the possible interactions.

Wind driven surface currents are strongest in shallow water. In semi-enclosed coastal waters the interaction of tide wind stress and Coriolis forces can generate complex three-dimensional circulations where neither the surface current nor the return flows are in line with the wind.

Mud transport

A striking phenomenon in turbid estuaries in the UK is the larger depletion of dissolved oxygen on spring tides compared with neap tides. This is despite the greater dilution afforded during spring tides.

Observations in the Usk and Parrett estuaries in the later 1970s indicated the importance of this phenomenon.[2] The relationship between suspended solids and dissolved oxygen is well illustrated in the Parrett estuary (Fig.3). As bed material is eroded by strong tidal currents on the flood phase of a spring tide a significant amount of organic material, including natural detrital matter, is eroded along with the end products of decomposition occurring within the bed during preceding neap tides. This relationship between increased suspended solids during spring tides and pronounced oxygen sags is also found in the Mersey estuary, and in the Ely estuary in South Wales.

Some models attempt to incorporate the effect by use of an additional oxygen sink term assumed to be proportional to the tidal range or to suspended solids concentration. Such approaches do not permit the prediction of water quality in cases where the hydraulic and sediment regime of an estuary is changed by engineering works or where the input of particulate BOD is changed. An adequate simulation of the effect of particulate BOD requires not only that the flow is well simulated but also that the transport, deposition and erosion of particulate matter is handled realistically and also that degration within the bed is simulated.

Discussion

In order to model water quality in estuarine and coastal waters adequately it is necessary to ensure that the flow model is capable of reproducing all the significant hydrodynamic and physical processes. The decision as to which processes are, or are likely to be, significant in a particular application needs to be based on a full appreciation of the situation. Narrow well mixed estuaries can be readily simulated using a one-dimensional model. However the inclusion of a part tide barrage downstream of the limit of saline intrusion could well result in stratification occurring upstream. The depth of the halocline and its sharpness would depend on such things as river flow, tidal range as well as on any operational criterion. An adequate models must be able to simulate such variation if water quality is to be predicted with any accuracy.

The choice of model type requires the skill and experience of the modeller if the predictions are to be meaningful. The choice of a simplistic model to give an approximate answer is almost certain to be problematical if significant processes are ignored. Omission of relevant processes precludes any estimate of accuracy of predictions outside the conditions for which a model was calibrated. The use of a one-dimensional model to predict the effects of remedial measures in a stratified estuary or one with significant gravitational circulation may result in grossly optimistic or pessimistic conclusions and it may well not be possible to even guess which is the case.

One, two and three-dimensional models are all useful tools if used appropriately in the right situation. This paper has highlighted some major hydrodynamic and physical considerations which need to be taken into account when applying water quality models.

References

1. Odd, N.V.M. and Rodger, J.G. Vertical mixing in stratified tidal flows, *H. Hydr., Div.*, ASCE, **104**, (3), 337–351.
2. Welsh Water Authority (1981) The Usk Estuary; Volume 1, An assessment of sewage disposal facilities, Tidal Waters Report No. 81/7.
3. Maskell, J.M. (1985) Particulate BOD and oxygen balance, in: J.G. Wilson and W. Halcrow (eds), *Estuarine Management and Quality Assessment* Plenum Press, New York, pp. 51–60.

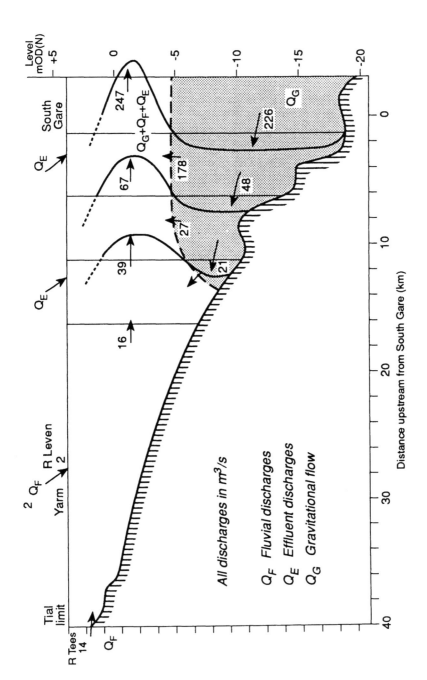

Figure 1 Typical gravitational circulation

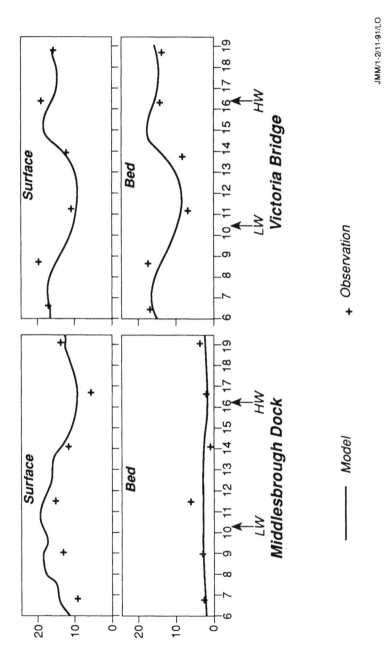

JMM/1-2/11-91/LO

Figure 2 Simulated and observed BOD concentrations (mg/l) in Tees Estuary, spring tide, low fluvial flow

117

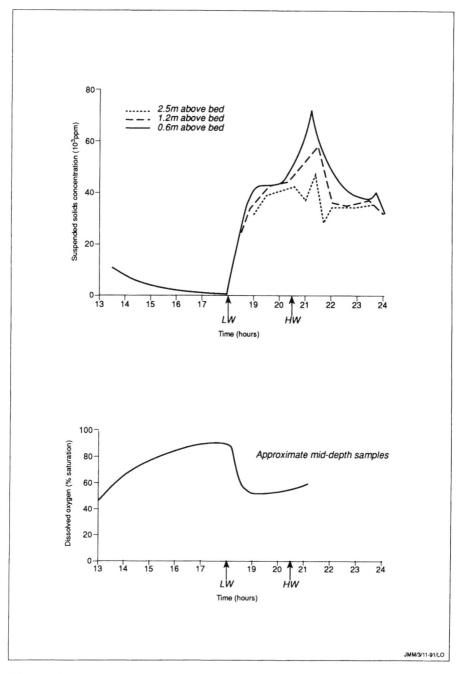

Figure 3 Effect of suspended solids on dissolved oxygen in Parrett Estuary

7 Water quality aspects of estuary modelling

J.I. Baird and K. Whitelaw

Introduction

The UK water industry is currently investing significantly in improving the quality of our rivers, estuaries and coasts. It is important to ensure that investment is targeted at those areas where most environmental benefit can be achieved, particularly in view of the large sums of money involved. Models are effective management tools and, as such, are vital components of any management strategy to bring about both short and long term improvements in water quality. The industry has long recognized the role of computer models and has been responsible for developing or promoting the development of a wide range of models, each tailored to particular requirements. The need for advanced models is set to continue in a climate where improving water quality is high on the agenda. This paper addresses those areas of estuary modelling which WRc perceive to be important over the next few years.

The Water Act 1989 allows the Secretary of State to set Statutory Water Quality Objectives (WQOs) for classified controlled waters including estuaries. These objectives are likely to include classification schemes. The National Rivers Authority (NRA) will play a principal role in defining the nature of an estuary classification scheme. It can be expected that the estuarine water quality classification adopted by the Secretary of State will be more objective than the current NWC classification which is loosely based around biological, chemical standards and aesthetic criteria. Whatever classification scheme is

119

finally chosen, there will be a clear role for models to provide assessments of estuary performance and to assist in the prioritization of improvement plans and defining monitoring strategies.

In modelling water quality in estuaries the issues which are considered to be important are:

- Conceptualization of the estuarine system.
- Characterization of the polluting loads.
- Framework within which Water Quality Standards (WQOs) are set.

The estuarine system

Both flow and chemical processes in estuaries are complex. The continually reversing tidal currents with often small residual currents mean that pollutants, while being widely dispersed within the estuary can have a long retention time. Long term accumulation of contaminants such as heavy metals and organic compounds can occur in the sediments. Density driven currents resulting from the interaction of freshwater flows to the estuary with the more dense saline coastal waters can complicate the tidal dynamics. Add to all of this a semi-diurnal pattern of significant changes in tidal levels and mudflat exposure and the modeller is presented with a far from easy task of simulating a complex environment by means of software code which incorporates advanced numerical methods to achieve accurate predictions.

Discharges

Effluent discharges to estuaries often have only preliminary or primary treatment and the dispersive tidal currents are relied upon to prevent reductions in dissolved oxygen levels. Many of these discharges are intermittent or seasonally varying in quality and quantity.

Water Quality Standards

The NWC classification for estuaries at present relates to biological, chemical and aesthetic criteria. It is probable that WQSs in the future will be more sophisticated and will reflect for example, the increasing concern over the potential toxicity of contaminants within the framework of EC directives. This requirement will generate a need for models which simulate these more complex processes.

In summary, there is a challenge for the modelling community to develop models which can simulate the numerous estuarine processes and the fate of both simple and complex contaminants introduced to the estuary.

Areas of development

Modelling framework

Many in the scientific community argue the case that the physical, chemical and biological processes in estuaries (or other waters for that matter) are poorly understood and investment in research is essential. This is indeed true, but what is often overlooked is the extent to which models address the particular requirements of the end users. If models are to become accepted as tools to be used in house, then the framework within which these models operate must be considered alongside their process capabilities.

Two aspects of this framework are software quality and the use of statistical techniques.

Software quality   Experience suggests that models which expose the user to selecting an array of complex coefficients, require intensive data manipulation, give poor graphical presentations, have minimal supporting documentation and consist of generally poor quality software, are unlikely to become widely used by water industry staff. Often the quality of the software is addressed as an afterthought. This area of model development is crucial if models are to be used successfully as routine management and planning tools by the regulatory authorities and dischargers. The alternative is for the industry to continue to rely on specialist skills found in consulting and research organizations for running the models.

Stochastic methods   It is recognized that all discharges have some measure of variability in both quality and quantity. The variability is often significant and needs to be considered in any realistic simulation of the impact of the effluent on the receiving water. River models such as TOMCAT and SIMCAT, which are used for consenting discharges, rely on stochastic methods to relate the impact of variable effluent loadings on river water quality. This is done by representing the effluent load by a frequency distribution and by taking advantage of the unidirectional flow of rivers. These models adopt a simplistic approach to process modelling using the argument that the sampling errors contained within the data outweigh errors which may arise due to an inadequate description of environmental processes within the model. In a similar way estuary models will need to consider the variability of input loads and provide statistical information with which consents can be determined, since WQSs are likely to be based on percentiles or averages. However, modelling these systems in this way is a more complex process and detailed hydrodynamic and water quality models will be required.

In addition to statistical methods, there still remains a requirement for models to consider particular events, such as periods of high temperature, or low flow, when water quality may be significantly impaired. This area of modelling discrete events and considering joint probabilities of input data to estuary models is being addressed by WRc in a NRA funded project. All the estuary models developed for the NRA by WRc in the past two years such as the Thames Tideway, Colne, Crouch, Roach and, more recently, the Stour and Orwell, have included a stochastic shell around the model to incorporate the variability of discharge loads. A simplified flow chart describes the approach adopted which has been named QUESTS (Figure 1).

Process modelling

It was recognized earlier that estuarine processes are complex and have a major impact on the behaviour and fate of pollutants. Our steadily improving knowledge of these processes needs to be incorporated into our models in order to improve their predictive capabilities. This applies to each of the physical, chemical and biological components of the ecosystem. The modelling process can also be used to identify those areas in which further research should be undertaken in order to improve our predictions and, hence, our management of the estuary environment. The processes for which models are required are shown in Figure 2.

Physical processes   Figure 1 showed that the tidal level characteristics for any particular day are input to the model as a seaward boundary condition. This stimulation of actual tidal conditions, particularly the spring/neap tidal cycle, is seen as important to the modelling of water quality because of the role of residual currents and to generate realistic statistical information. The QUESTS model, as applied to those estuaries named earlier, incorporates tidal harmonic relationships similar to those used in the generation of Admiralty Tide Tables. Episodic events such as storm surges are not included in the model, although the impact of storm-generated waves on the long term accumulation of sediments may have implications for sediment borne contaminants.

Chemical processes At present, most models consider the standard determinands which contribute to the dissolved oxygen balance in the estuary. These processes are:

- Degradation of organic carbon compounds (BOD).
- Degradation of nitrogen compounds (Organic nitrogen, ammonia, nitrite, nitrate).
- Photosynthesis (algal growth/respiration).
- Bed respiration/resuspension of organic sediments.
- Reaeration.

The number of determinands solved by existing models in determining dissolved oxygen levels varies between 5 and 15 depending on the degree of sophistication adopted. Coefficients used in the associated equations are reasonably well established, although there is always an ongoing requirement to improve decay rates and reaction rate constants as more data is obtained: for example, our understanding of nitrification rates in estuaries is improving as a result of the work being undertaken by WRc on behalf of the NRA.

The main developments in modelling the more complex chemical determinands are likely to be in the following areas:

- *Eutrophication* The urban waste water directive identifies water bodies as *sensitive* if they are shown either to be eutrophic now or likely to become eutrophic in the near future. Understanding the eutrophication process and the effects of phosphorus and nitrogen on algal behaviour is important. It is desirable that reduced eutrophic status can be demonstrated for any scheme involving nutrient removal.
- *Metals/organics* Concerns over the impact of List I and II substances to estuaries will be formalized increasingly in directives and standards. These contaminants often have an affinity for suspended material through adsorption. Hence their fate becomes closely linked to the erosion and deposition characteristics of the sediment. The complex partitioning relationships are only now beginning to be understood and conceptualized in computer models.

    Ongoing experiment work at WRc on behalf of the NRA and DoE continues to improve our understanding of metal behaviour. There are clearly developments still to be made in improving our ability to model speciation processes, particularly with regard to the fate and behaviour of the more complex organics.

Biological processes The UK approach to water quality management is to manage and control discharges on the basis of Environmental Quality Standards. These standards are currently expressed as concentrations of chemical determinands, the threshold levels of which have been set using toxicity information and an understanding of the potential hazards associated with the determinand in question. Increasing attention is now being given to standards which include biological components, such as the requirement that particular indigenous species should exist in a water body for that water body to achieve a high classification. In addition an EC directive is being developed based on the concept of ecological quality. This will focus on the status of planktonic, invertebrate and fish communities. This need to understand and determine impact of effluents on the estuarine ecosystem will be reflected in a requirement for biological and ecological models. With these tools the regulatory authorities and dischargers will be able to predict the impact of activities on estuarine ecosystems using input data from hydrodynamic and water quality.

Models are currently being developed at WRc which predict the body burden of a contaminant in selected species from exposure rates, both in the water and in the diet. The body burdens are related to sub-lethal (growth and reproduction) effects through comparisons with appropriate body-burden/toxicity relationships. This work is at an early stage of development and a great deal needs to be understood about the response of the estuarine community before models can be used to predict with confidence the effect of a particular compound on a diverse ecosystem.

Conclusions

Water Quality Modelling is seen to be an essential tool for the effective management of estuarine systems and for the design and optimization of discharge regimes. In this age of environmental responsibility, the model can be used to help the decision making process so that environmental objectives can be met, adequate safety margins established at a realistic cost and priority measures identified. Modelling is being driven by these needs and this is reflected in the observation that a model of today is a different animal from one of 5 years ago. Estuary water quality models are needed to address both the problem of consenting discharges which are variable by nature and that of simulating discrete events. More sophisticated process models are required which incorporate our improved knowledge of the physical, chemical processes in an estuary. At the same time our understanding of biological processes is growing and effort is now being directed to incorporating this knowledge into simple models. With the growing importance of biologically-based WQSs, the development and linking of biological/ecosystem models with chemical water quality models is seen as an area of rapid development.

Finally, it is important that the end-user has a model which is easy to use. In order to meet this requirement, it is preferable that the model has a user-friendly interface, good graphics and is robust. Models will still need to be developed by organizations with specialist skills. However, the end product must be capable of being used by people who do not possess these skills, but who have an understanding of the benefits and limitations of models. It is therefore essential that good software programming practice becomes an integral part of model development.

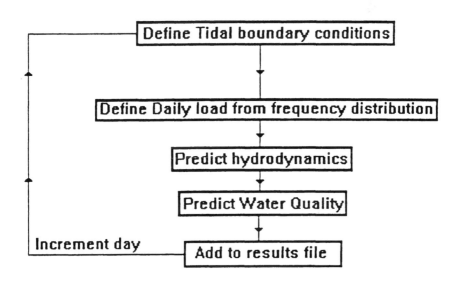

Figure 1   Schematic representation of QUESTS

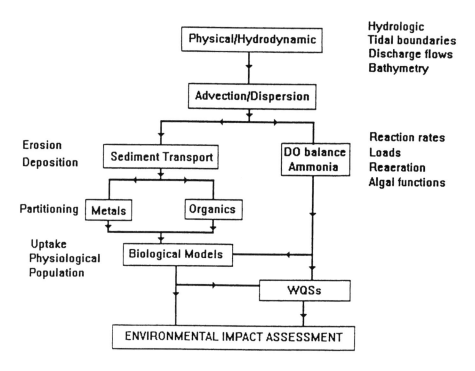

Figure 2   Impact assessment through models in an estuarine environment

8 Mathematical models and engineering design

Graham Thompson

Abstract

The role of mathematical models in engineering design is no longer that of simply automating techniques that were previously carried out manually. Throughout industry models are now becoming accepted as one of the main decision support systems to managers. This is certainly the case in engineering design for managing the environment. We are rapidly moving into the age of expert systems and hydroinformatics, where the primary aim of the model is a decision support stem. In this paper the role of the models in modern practice is reviewed and illustrated by case histories.

Key words: Coastal; Design; Management; Modelling; River; Water quality.

Introduction

Models, be they mathematical, physical or conceptual, are interactive decision support tools used by modern managers and designers. Models are no longer an end product themselves, but their continual evolvement and improving reliability permits decisions to be made with more confidence and hence allows more efficient use of our natural resources.

Models and the management process

Civil engineering design can be considered as the overall management of our natural resources, as is stated in the ICE charter. The basic management process is to collect and assess facts so that informed decisions can be made on future strategy and the actions to be taken. These actions, when implemented, are monitored creating a feedback loop that allows further corrective action to be taken as necessary.

In order to attain sustainable water pollution control this management loop breaks down into the following steps:

- Monitor performance to assess need for action.
- Identification of potential problem.
- Determine appropriate action or scheme.
- Detail design and implementation of the scheme.
- Continue to monitor performance to assess need for further action.

Models have a part to play in decision support in all aspects of this basic management loop.

Monitoring and identification

To determine whether or not a potential problem exists requires both:

- a knowledge of existing and future potential conditions; and
- a set of criteria against which to judge the acceptability of those conditions.

Testing the compliance of conditions against criteria is the main means of determining whether action is required or not. Non compliance may arise due to one or more of the following:

- increased loading on the system.
- changes to the system.
- changes in the criteria.

Water quality modelling has a role to play at this problem identification stage in both the infilling of information on the existing system and also by providing estimates of the future conditions. It is often the case that the data available on the existing condition of a river, lake or coastal area is sparse and limits our appreciation of the processes occurring within that water body. By carrying out a parametric study, using a model, at this stage it is possible to identify the significant features of the system and hence improve understanding so that decisions on whether a real problem exists can be made from a position of knowledge rather than based on what might have been incomplete or anomalous data.

To predict future conditions and allow strategic planning some form of model will be required. As a starting point such a model should be capable of representing existing conditions, although this is not always the case. If the future conditions are to be radically different from the present conditions then there is little point spending a lot of resources on fine tuning a model to represent the existing conditions if the controlling processes in the new system will be different. For instance there is little point in simulating in detail the existing conditions within an estuary if the future scheme is likely to be to collect and dispose of all effluent through a long sea outfall well away from the estuary. The model may well be useful to investigate partial treatment options, or phasing of implementation of the scheme, but the expense of an elaborate study could be avoided if the long sea outfall was the only main choice. The main purpose of modelling at this stage is to test the overall capacity of the system to accommodate the foreseen changes.

Determining appropriate action or scheme

A major use of models occurs within the second stage of the process, i.e. in deciding on what action to take and what scheme to build. This stage is commonly called the masterplan and subsequently the feasibility stage. In these stages the 'no action' scenario will usually have been evaluated to quantify necessity and the benefits of a scheme as part of its justification.

The management team's main role at this stage is, for a full range of possible options, to:

- Identify the totality of the impact of the scheme, i.e. the scoping study.
- for each impact identify:
  - the controlling processes that will effect the decision;
  - the appropriate models for investigating that process, if any is needed or available;
- evaluate each impact, both the effect of the project on the environment and the effect of the environment upon the project;
- determine the overall benefit of the scheme taking into account the result of evaluation of each impact, the reliability of each evaluation and also other relevant considerations such as feasibility of construction, economics and politics.

These evaluations are carried out for each of the originally envisaged schemes and any mutations that emerge during the study. The detail of the investigations will almost certainly vary between options depending on the initial indications of the viability of each. The conclusion of this stage is a single scheme, which will probably have been agreed through public consultation, to take forward for detailed design.

The models to evaluate these impacts will vary as the project develops. For instance, the start of a project requires a broad brush model that can rapidly evaluate the strategic options. This may be a regional model of a coastal area, or a full catchment model. It need not necessarily consider the processes involved in detail. Often mass balance techniques and mixing pot water quality models can be sufficient to identify the potential of the main options. However as schemes become defined in more detail it is necessary to look in detail at the actual processes occurring taking into account not only the main processes, but also the secondary processes and their interactions.

Detail design and implementation

During this stage modelling is often very detailed. It is no longer only necessary to use models to compare schemes but to give quantitative information on the actual design required. The large complex models of water bodies used in the previous scheme determination stage often have to be supplemented by models yielding results at a finer level of detail, although even then the results have to be tempered with judgement before the design criteria and details are finalized.

As an example, the type of models used at this stage for a sea outfall project may include diffuser hydraulics and plume development models, surge analysis of the outfall and headworks, detailed models of the headworks themselves and the contributing catchment.

Monitoring and performance

There is a natural tendency to think that, once a scheme has been implemented, it is unnecessary to continue monitoring with the same vigour as previously. However the potential benefit of a scheme can often be optimized by post implementation monitoring and to adapt the operation rules of the scheme according to observations. Such post project appraisal is an important part of any scheme and should be built into the capital costs of the scheme. It may reveal that the scheme is sensitive to particular environmental conditions or social pressures that were not considered in detail during design, but which could be accommodated by minor changes.

In the past the use of models in such studies has been limited, primarily due to the expense of setting up new models for a re-evaluation. However now that more owners are either carrying out the modelling work themselves or having the models mounted on their own systems at the project design stage it should become easier in the future to integrate the use of models with the monitoring programme and to have them available for use in post project appraisals and future problem identification studies.

Available models

The previous chapters in this book have introduced many key features embodied in the latest types of models available. They have extolled the virtues of a comprehensive assortment of models capable of predicting firstly the water movement under a wide range of physical conditions and subsequently the variation of the quality of the water body for differing pollution loads, taking into account the physical, chemical and biological processes. The models described have covered a range of scales, both in time and space.

There are those models that work with time scales in the order of months and distance steps in the order of many kilometres. These are necessary to predict the seasonal variation of salinity in major estuaries or nutrients and radionuclide in large coastal seas.

At the other end of the spectrum there are those models which seek to build up a global picture by considering the fine detail of the component parts of turbulence structure and diffusion processes that have time scales in the order of milli-seconds and length scales in the order of millimetres.

This range of models is complimentary, not exclusive. Each model is appropriate for its own scope of problems. Models are not reality. The advances in the capabilities in modelling that have been discussed in previous papers are essential to the improved efficiency with which we use our natural resources. Research being carried out will underpin the models of the next decade or more, providing more reliable and more comprehensive tools. But models will not in the foreseeable future be able to combine the infinitely small time and distance scales with the ability to integrate accurately these effects over minutes, days, months or even centuries, and over metres, kilometres and global seas, as nature itself does. All models are by definition only representations of reality. Furthermore they only represent reality as seen from a particular perspective. For example a model capable of predicting the initial dilution at an outfall is not appropriate for predicting regional water quality and vice versa. The scales, the assumptions and the limitations of each model are different.

What is required for effective environmental management

The prima facie decision for the environmental manager, and this is often the engineer, is not the detail of which model to choose for his application, but the overall strategy to follow and scheme details such as whether the outfall should be 1 km long or 1.5 km long , or whether he should or should not object to a proposal for a 1500 MW combined cycle power plant in his region.

The major follies of our time are rarely due to miscalculation of the detail, but more often the lack of appreciation of the totality of the effect. For instance High Aswan totally fulfils its purpose to regulate the Nile, but the persistence of bilharzia in the reservoir and the lack of sediment feeding the delta, which is resulting in the wide spread erosion of the Alexandria coast, are problems that were never conceived of, let alone evaluated at the project formulation stage. I'm sure that we can all think of cases closer to home.

It is the responsibility of the environmental manager to foresee the total range of issues that will be effected by his or her decision, this can not be done by models, although the emerging breed of expert and hydroinformatics systems will go some way in this direction. Having identified the issues these must be translated into a number of studies of individual processes, and each approached with the appropriate tools. Subsequently the manager has to assemble the outcome from each study into a coherent justification for the course of action that will eventually be pursued. Furthermore the justification for the selected scheme has to be agreed with non technical audiences, who will include the project funders, politicians and the ever more demanding public.

The skill of the management team is their ability to identify the totality of the potential impact of the scheme and furthermore recognize those aspects that must be quantified with confidence, those that can be quantified with less confidence and those that need to be taken into account, but not quantified. The management team must have the ability to select the most appropriate approach, including models where necessary, for each area of investigation. This is a two fold process.

The management team has to be fully familiar with the physical, chemical and biological processes that will have a significant impact upon the ultimate decision. The investigations must be restricted to these areas of significance, otherwise costs will rapidly escalate for no perceptible improvement of quality of the conclusion. However, the detail of the significant processes can not always be defined in advance and therefore a wider initial study has to be carried out with key issues being investigated at increasing levels of detail as the project evolves.

The management team has also to be aware of the limitations of each method of approach and model types so that they know the confidence to place in the results. This is just as important as knowing the actual results. As long as tolerance limits are known then these can be allowed for.

By selecting the correct model and by simulating conditions within the validity of the model, the results can be expected to be reliable for a full range of 'what if' questions. The model will allow a range of conditions to be considered and the most appropriate course of action determined, within the relevant constraints and criteria. Furthermore the use of advanced graphics software and digital terrain models can allow the simplicity of the more complex aspects of the justification to be presented to non technical audiences.

Examples

In the rest of this paper I have used examples from recent projects where we have used a number of models to assist in the decision making process at different stages of the project.

River Yangtze Chong Qing

Chong Qing is a major industrial municipality with a population of 13 million, at the confluence of the rivers Yangtze and Jailing in central China. The rivers provide both the water supply and the main route for effluent disposal. Although there is a sewer system much of the urban and industrial effluent enters the rivers without any form of treatment. In the more congested areas of the city it is common to see alternate water intakes and outfalls along the banks of the rivers.

Although Chong Qing is well inland, the rivers are still large in European terms. The Yangtze has a dry weather flow of 2,000 m^3/s whilst flows of 20,000 m^3/s are exceeded typically on 15 days per year. The flows in the Jailing are approximately 1/3 of the flows in the Yangtze. A reach length of 220 km of the Yangtze lies within the Municipality, the figure for the Jailing being 140 km. Given these large flows and long distances there is opportunity for considerable dilution of effluent and self purification of the rivers as they flow through the Municipality. However, present effluent quantities during low flows in the rivers are 1% and 5% of the river flows for the Yangtze and Jailing respectively. In addition the rivers already have high pollution levels and bacteria counts when they enter the Municipality.

The two main questions to be addressed were:

What is the long term strategy that should be adopted?
What could be done to improve the quality of water at existing intakes, in the short term?

To address the long term strategy we used the one dimensional planning model, QUALITY, that had previously been developed by Binnie & Partners. This represented the water quality along the full length of both rivers within the Municipality. The model considered BOD, DO, ammonia, phenol, cyanide, coliforms, nitrate and heavy metals including chromium, copper, cadmium, zinc and lead.

The model was installed on computers in the offices of the Authorities in Chong Qing and training given on all aspects of pollution control. The model of the existing system gave the authorities a clear indication of the relative importance of the industrial loads, the domestic loads, the aerial pollution that is washed back into the river, the upstream pollution and the self-purification capacity of the rivers. With this understanding they have formulated a long term plan that seeks to reduce the aerial component of the polluting sources, to discharge effluent downstream of the main conurbation and to selectively treat industrial effluent on site. The Authorities are also now fully aware of the limited impact that this will have on the more conservative pollutants that originate outside the municipality.

The rivers at Chong Qing are wide. The Yangtze is some 500 m whilst the Jailing is some 300 m. The water quality at the intakes can be improved by extending the outfalls to avoid direct contamination of the intakes. To study the effectiveness of this proposal a 2-dimensional model was set up. The flow patterns vary rapidly along the river and hence to represent them with a regular grid hydrodynamic model would have required a fine spatial resolution that would have been computationally prohibitive on the PC based systems available. The solution adopted was to use a streamtube method developed by Holly[1] for the Missouri. This has been coded by Binnie & Partners into a model that is referred to as DISPERS. In this method the channel section is divided into a number of streamtubes, determined from field measurement of flows across the section. Within each streamtube the longitudinal flow is constant. Concentrations are calculated along each streamtube using a one dimensional analysis, but taking into account the widthwise dispersion of concentration. This method proved to be computationally robust and readily applied with limited computer power. In addition it accurately took into account the local variations in flow profile across the river. Dye dispersion tests were carried out in the river to establish realistic lateral diffusion coefficients.

The model is now being used by the Chinese Authorities to determine the extension required for individual outfalls near major intakes. Trials indicated that whilst the derivation of general rules was possible most cases were site specific and dependant upon local flow patterns so each major intake is being considered separately.

Within this project the strategy model QUALITY was used to collate information on the existing conditions in the rivers and to determine the significance of the main contributing factors to the present adverse water quality. This allowed the managers to clarify the key issues and to hence determine a long term strategy that will result in a marked improvement in water quality particularly in the reaches were water intakes abound.

The model DISPERS was used to determine in detail the arrangements that should be adopted for outfalls and intakes in the river to minimize the direct connection between the two.

Victoria Harbour Hong Kong

In January 1988 Binnie & Partners were appointed by the Hong Kong Government to prepare water movement and water quality models of the western approaches to the main harbour in Hong Kong. This work was carried out with Hydraulics Research Limited (HR), Water Research Centre (WRc) and EGS as subconsultants.

It was foreseen by the Hong Kong Government that increased reclamation of land on the fringes of Victoria Harbour would have an impact on the water quality of the surrounding territorial waters. The requirement of the brief was to set up a comprehensive suite of numerical models and a physical model to allow the impact of the proposed developments to be rigorously tested.

In order to understand the scope of the modelling being asked for it is necessary to understand something of the oceanographic conditions in Hong Kong. The colony consists of the main island of Hong Kong, the area known as Kowloon on mainland China and numerous small islands within a sea area of approximately 60 km by 45 km. Hong Kong is situated at the mouth of the Pearl River estuary. During the wet season stratification can be pronounced in the tidal waters around Hong Kong although it is not evident during the rest of the year. In addition to the complications of changing stratified regimes, the oceanographic currents near Hong Kong change direction seasonally complicating the boundary conditions for the models.

135

The main model suite consisted of:

- depth average intertidal water movement and water quality models which simulated the dry season conditions. The water movement model covered most of the Pearl estuary at a grid size of 750 m, which reduced to 250 m within HK territorial waters. The water quality model covered the territorial waters only.
- In order to simulate the wet season conditions 2 layer intertidal models of the territorial waters were established. These simulated water movement, salinity, water quality and sediment transport.

These models were able to evaluate many of the effects of proposed developments and determine the extent to which flows are diverted away from Victoria Harbour as large land reclamations take place. It is of interest to note that in some circumstances the depth average model gave worse water quality conditions, i.e. during the dry season. However, in other cases the stratified condition during the wet season gave worse conditions as pollutants became trapped below the surface layer and were unable to decay or become re-oxygenated. Both wet and dry season models were necessary to evaluate most proposed developments.

In addition to the intertidal planning models it was necessary to develop a seasonal water quality model that was driven by residual currents set up by longitudinal density differences. This model did not consider the intertidal oscillation of water along the coastline and therefore could use long time steps so that simulations covering a year or more were possible. This model used large spatial elements in plan, some being as big as 10 km square, and in depth it had a reasonably fine resolution with 5 layers. This model was found to be invaluable in carrying out the overall assessments of ammonia, BOD and nitrates as they changed seasonally through the year, partially as a result of the Pearl river flows and partially a result of the seasonality of the pollution loadings and processes.

The above mentioned models provide advice on the general policies to follow. However, it was also necessary to produce models that allowed the detail of schemes to be defined. For this a system was established for using the results of the depth average models to drive local models at fine scale, usually 83 m but sometimes less. These finer grid models could be set up in any orientation to evaluate the hydraulic and water quality aspects of smaller features, such as the small harbours used as typhoon shelters and also to model the detail of individual reclamations.

In addition in order to evaluate the effectiveness of the many sewerage outfall models around Hong Kong and to assess the effect of the proposed large offshore outfall a further bacteriological random walk outfall model was developed. Again this can be set up to work anywhere within the coverage of the depth average model.

As well as the above mentioned models, wave prediction models were added to the suite and an overall user friendly shell and data archive system was added. In order to set up the models an extensive data collection programme was also carried out. The computer programs were transferred to the Hong Kong Government who now use them extensively. Furthermore a large purpose built laboratory and physical model was provided in Hong Kong to test out the local flow profiles in the vicinity of some of the reclamations and to increase public awareness and confidence in the work.

It can be seen from this list of models that in order to evaluate the totality of the impact of the major schemes envisaged for Hong Kong the decision makers realized the importance of establishing data acquisition and modelling facilities at an early stage so that not only could their knowledge of the existing system be improved but so that they have the predictive capacity to make informed judgements on new proposals as they arise.

Fraserburgh Long Sea Outfall

It is the intention of Grampian Regional Council to construct a long sea outfall at Fraserburgh, in north east Scotland. As part of the feasibility study a coastal model was set up to determine the potential for compliance of outfalls at varying locations. As with other such outfall studies we used Bradford University's DIVAST model. The model was driven from POL's 9 km grid model of the North Sea. A large part of the DIVAST model was established with a 3 km grid, but within a tidal excursion on either side of the outfall locations this grid was reduced to 333 m. The model was set up to cover a large sea area of 140 km by 100 km in order to simulate the complex flow conditions that exist. During the flood tide the flows initially flow eastward towards the Moray Firth, but as the tide advances the currents swing easterly and flow towards the North Sea. As the tide changes the current swings round to flow in a westerly direction, in which direction it continues until well into the flood tide. These conditions were represented well in the model, but only when the driving boundary conditions were removed sufficiently far from the area of interest, so that the sea bed topography could exert its full influence on the tide.

The model was set up using the Moss digital Ground model to both generate the 20,000 bathymetry points and also to act as the basis for the graphical presentation of results. This was particularly useful as it allowed the current vectors to be superimposed on colour contoured plots of the sea bed topography allowing the influence of the sea bed contours to be readily observed.

The main criteria for compliance was the faecal coliform levels. The model produced these results at various states of the tide for different wind and tidal conditions and for different outfall locations. Colour plan view plots of these results were produced with the colour coding corresponding to exceedence of or compliance with both the EC mandatory levels and the guideline levels. These plots allowed rapid evaluation of the acceptability of each outfall location. In addition plume centreline plots were produced for each outfall to allow visualization of the actual spread and movement of the plume in the vicinity of the diffuser.

The model allowed the range of potential outfall sites to be compared. Sensitivity analysis allowed the performance of each outfall site to be tested for a range of tidal, meteorological and effluent loading conditions. Sensitivity to die off rate was tested, including very slow die-off rates which would be representative of conditions through the night or to the more persistent viruses. Having carried out these tests the final chosen outfall location could be relied upon to be robust to a wide range of physical conditions.

This application is a classic example of a model being used at the feasibility stage of a project to compare a wide range of options and conditions and to present the results in a clear form that can be readily understood by non technical audiences.

Humber Estuary Tidal Power

This final example again illustrates the use of DIVAST, but this time with the inclusion of fully operational tidal power barrages within the area being modelled. The modelling was carried out recently for the initial pre-feasibility stages of this project. The model was used to consider a wide range of technical features of two barrage sites, one being across the mouth from Spurn Head, the other being half way between Immingham and Hull. The model included the full estuary from some distance out into the North sea to Goole on the Ouse and Newark Weir on the Trent using a 300 m grid. The model was used to consider the effect of each barrage site on not only the tidal flow within the estuary but also:

- the changes in sediment regime;
- the implications on water levels and hence land drainage and wildlife habitat;
- the effect upon tidal excursion and the dispersion and dilution of

effluent from existing outfalls both upstream and downstream of the barrage sites.

Furthermore the model was also used to determine the available power output from the two barrages for differing numbers of turbines and sluices taking into account the fully hydrodynamic effect upon the estuary.

In this application the model was used to determine the implication of overall changes to the physical system itself rather than changes to loadings on a system that stays essentially unchanged as is commonly the case in most water quality models.

Conclusions

Water quality modelling is not an end in itself. It does however provide the means of obtaining information on which to base decision and designs that are environmentally sustainable. Models have a role to play in all aspects of a project from the initial monitoring and problem identification stage through the scheme definition, detail design and implementation stage and in the operational, monitoring and post project appraisal stages.

Every model has its limits. Models are only a representation of reality and each type of model has different limitations and approximations. Models are however the most powerful tool available for clarifying what are the significant processes occurring in a natural system and for predicting the effects of change to a system. The nature of the decisions that have to be made change as a project develops and so do the model requirements. Therefore either the model has to evolve with the project or different models have to be used.

The skill of the team in charge of the overall design of a project is their ability to foresee the totality of the potential impacts of the scheme on the environment. They must have the ability to use the most appropriate tools, including models, to evaluate each impact and have the confidence to modify the scheme if necessary. Such decisions are based on an understanding of the accuracy of the predictions and also the relative importance of each impact in terms of the wider social economic and political objectives of the project.

Reference

1. Holly, F.M. Jr (1975) Two-Dimensional Mass Dispersion in Rivers, Hydrology Papers No. 78, Colorado State University, Fort Collins, Colorado.

Printed and bound by CPI Group (UK) Ltd, Croydon, CR0 4YY

21/10/2024

01777087-0019